Springer Theses

Recognizing Outstanding Ph.D. Research

Aims and Scope

The series "Springer Theses" brings together a selection of the very best Ph.D. theses from around the world and across the physical sciences. Nominated and endorsed by two recognized specialists, each published volume has been selected for its scientific excellence and the high impact of its contents for the pertinent field of research. For greater accessibility to non-specialists, the published versions include an extended introduction, as well as a foreword by the student's supervisor explaining the special relevance of the work for the field. As a whole, the series will provide a valuable resource both for newcomers to the research fields described, and for other scientists seeking detailed background information on special questions. Finally, it provides an accredited documentation of the valuable contributions made by today's younger generation of scientists.

Theses are accepted into the series by invited nomination only and must fulfill all of the following criteria

- They must be written in good English.
- The topic should fall within the confines of Chemistry, Physics, Earth Sciences, Engineering and related interdisciplinary fields such as Materials, Nanoscience, Chemical Engineering, Complex Systems and Biophysics.
- The work reported in the thesis must represent a significant scientific advance.
- If the thesis includes previously published material, permission to reproduce this must be gained from the respective copyright holder.
- They must have been examined and passed during the 12 months prior to nomination.
- Each thesis should include a foreword by the supervisor outlining the significance of its content.
- The theses should have a clearly defined structure including an introduction accessible to scientists not expert in that particular field.

More information about this series at http://www.springer.com/series/8790

Florian Willomitzer

Single-Shot 3D Sensing Close to Physical Limits and Information Limits

Doctoral Thesis accepted by
the University Erlangen-Nürnberg, Erlangen,
Germany

Springer

Author
Dr. Florian Willomitzer
Department of Electrical Engineering
and Computer Science
Northwestern University
Evanston, IL, USA

Supervisor
Prof. Dr. Gerd Häusler
Institute of Optics, Information
and Photonics
University Erlangen-Nürnberg
Erlangen, Germany

Date of Submission: June 2, 2017

Date of Examination: August 11, 2017

Presider of Doctoral Affairs Committee: Prof. Dr. Georg Kreimer

Presider of Examination Board: Prof. Dr. Hanno Sahlmann

First Reviewer: Prof. Dr. Gerd Häusler

Second Reviewer: Prof. Dr. Richard Kowarschik

Third Reviewer: Prof. Mitsuo Takeda, Ph.D

ISSN 2190-5053 ISSN 2190-5061 (electronic)
Springer Theses
ISBN 978-3-030-10903-5 ISBN 978-3-030-10904-2 (eBook)
https://doi.org/10.1007/978-3-030-10904-2

Library of Congress Control Number: 2018966403

This Springer imprint is published by the registered company Springer Nature Switzerland AG.
The registered company address is: Gewerbestrasse 11, 6330 Cham, Switzerland

Supervisor's Foreword

Optical 3D sensors are ubiquitous in industry, medicine, artistry, and virtual reality. Surprisingly, the great majority of 3D sensors is useful only for static applications. In spite of significant demand, a 'high quality', single-shot 3D camera, capable of acquiring the shape of dynamic objects, is not available. Why is that? Mr. Willomitzer's thesis is essentially about this question and a novel '3D movie camera' that presents an answer.

The number of 'different' commercial 3D sensors is immense. Designers have a rich choice from a set of essential features—for the illumination, for the interaction of light with the object, and for the type of exploited information. To number just a few: Temporal/spatial coherence/incoherence, monochromaticity/broad band, structured/unstructured, continuous/pulsed, polarized/unpolarized, amplitude, phase, intensity, time of flight,… We counted the number of permutations, ending up at about 8,000 different 3D sensors.

There is need to bring some order to this plethora. This order might help for another problem as well: How to quantitatively compare 3D sensors in terms of output and efficiency? Eventually we want to know, is an advertised sensor really 'optimal'?

Mr. Willomitzer's thesis approaches these problems by looking at the limits of optical 3D sensors, which are given by nature and information theory, and how to reach these limits. Eventually, he presents a 'single-shot 3D movie camera' with the maximum possible data density and the best possible precision.

At the University of Erlangen, the limits given by physics have been investigated for many years, and several 'optimal' sensors found their way into industry and other fields. The concept of 'looking for the limits' has proved to be quite useful (possibly not only for opticists) because:

- Knowledge about the physical limits allows for the optimal sensor design.
- Limits often appear as uncertainty products, offering the opportunity to bargain with nature by sacrificing less important information for a better measurement of the desired information.

- Knowledge of the limits is of significant practical importance: we can avoid unnecessary technical effort by knowing that a device already reaches the physical limit of precision. We can also evaluate our competitor's data sheets with respect to 'impossible' features.

But there is one important question where knowledge of physical limits does not help: How much effort must be invested to achieve certain specifications, or (and this is not at all an un-physical question)—how much will a sensor cost? Surprisingly, help comes from information theory.

An optical sensor can be considered as a communication system. A 3D sensor delivers a number of uncorrelated 3D points with a signal-to-noise ratio SNR, corresponding to '$A[bit]$' of information. In order to acquire an amount A of 3D information, much more information B in form of 2D raw data has to be captured. B is always larger, sometimes much larger than A.

In other words, optical 3D sensors are inefficient! Why are they inefficient? Where is the limit of the efficiency? How can we reach the limit? How can we exploit this knowledge for better sensors? These are the questions that Mr. Willomitzer attacks in his thesis. The questions are theoretically interesting and significant to the every day life of the optical metrologist. Two issues are of major importance:

- First, can we reduce 'costs' (measuring time, technology, footprint, …) by more efficient sensors?

- Second, and this is the key issue addressed by Mr. Willomitzer's work: can we make a single-shot 3D camera with a high density of 3D points, and without sacrificing precision? 'Single-shot' means that the 3D-information is acquired within one single video frame, like in a hologram.

When Mr. Willomitzer started his project, such a camera could not be found in the literature or among available optical 3D sensors. The deep reason is that it is principally impossible to acquire the distance of an object point, its local reflectivity and the bias illumination (three unknowns), from one single-video image. A single-video image simply does not contain sufficient information.

Due to this information deficit, available single-shot sensors need a workaround: they exploit encoded or sparse features (e.g., Microsoft Kinect One®). The encoding is necessary to deliver the so-called 'correspondence'. But obviously, the encoding devours space bandwidth, so single-shot sensors display a low-density 3D point cloud (although the low density is commonly obscured behind a posteriori interpolation).

How much space-bandwidth must be sacrificed? This question leads to the upper limit of the data density of a single-shot sensor. Mr. Willomitzer would answer by referring to holography and Fourier profilometry, where one-third of the available space-bandwidth is usable. He would go on to show that the sampling theorem in fact leads to the same limit.

Unfortunately, this knowledge does not teach how to reach the limit and it does not take into account the signal-to-noise ratio. The latter is necessary to qualify the efficiency of a sensor and to compare different sensors. For this purpose, Mr. Willomitzer borrows the 'channel capacity' from information theory. The channel capacity defines the maximal information content that can be transmitted by a channel (here by our sensor). The channel capacity is of high practical importance too, as it is connected with technical effort or, more trivial, with cost.

Mr. Willomitzer exploits the teachings of the channel capacity to introduce a 'channel efficiency', which he defines as the ratio of the maximal 3D information output versus the maximum 2D information content of the raw data. The efficiency of a number of current commercial 3D sensors is just a few percent. So there is much room for improvement!

The channel efficiency does not depend on the object, so it is an intrinsic property of the sensor and gives useful quantitative data about how efficiently the hardware investment is exploited and how different sensors compete in terms of maximum deliverable 3D information.

After these considerations, the question still remains how to reach the limits, specifically the maximum possible density, best efficiency, and maximum possible precision (best 3D-SNR). The problem has been investigated at Erlangen for many years, with a few work-around solutions, before the breakthrough idea was formulated by Mr. Willomitzer.

The solution will not be discussed in this foreword. Instead, it will be preserved as a secret revealed only to the adventurous reader, and attributed to the author of this thesis. Just a hint for the ambitious 'want-to-find-out-myself' readers: Although the method presented here is incoherent, it turns out—as was discovered after this thesis was written—that the concept for the best possible data density presented here is equivalent to the concept of two-wavelength holography, where one wavelength produces precision and the other wavelength (or the combination of the wavelengths) produces uniqueness.

The results in this thesis cannot be illustrated better than by the series of short '3D movies' (see e.g., tinyurl.com/3DMovCam12).[1] It should be emphasized that the movies display original data, neither low-pass filtered nor interpolated. This quality is possible because the 3D camera works close to physical and information theoretical limits. Each video frame contains the full 3D information about the visible part of the object surface. This makes it possible to adjust the perspective of the viewer, while watching the video (which is not possible in the so-called '3D cinema', which in fact displays just a series of stereo images). It should not be withheld that Mr. Willomitzer's 3D camera exploits a magic trick that at first seems impossible: in fact, it comprises four triangulation systems with only two video cameras.

Mr. Willomitzer's thesis includes outstanding contributions to optical metrology. He is, to my knowledge, the first who clearly formulated a measure for an 'information efficiency' of 3D cameras. In this thesis, he illuminates how to approach the

[1] Alternative link: tinyurl.com/3DCam-012.

information theoretical limit of data density and significantly improve channel efficiency. He solves the fundamental correspondence problem of triangulation without encoding. He is the first to present a single-shot 3D video camera that incoherently acquires the maximum amount of 3D information in one image. The 3D movie results display an unprecedented 3D quality, as the camera works close to the physical limits.

Mr. Willomitzer is not just a good scientist. He also worked as a part time physics teacher during his research at the University, just for fun. This clearly written thesis reflects his passion for explanation. The introductory chapters deliver the background to the fascinating world of optical 3D sensing and make the thesis a highly recommended read, not only for experts.

I am pleased that Springer is publishing this outstanding thesis.

Erlangen, Germany
November 2018

Prof. Gerd Häusler

Abstract

This thesis introduces a novel optical 3D sensor principle and its implementation: the *single-shot 3D movie camera*.

The camera is designed for the 3D measurement of macroscopic objects with scattering surfaces, e.g., human faces. It combines the acquisition of a *dense point cloud* displaying *physically limited lateral resolution and depth precision* together with a *single-shot ability*. 'Single-shot' means that *no temporal sequence* of exposures is exploited to generate the 3D point cloud. The approach is based on *multi-line triangulation*. Since, in contrast to other single-shot approaches, no space bandwidth is wasted by pattern codification, the 3D point cloud can be acquired with its maximal possible density: A 1-Megapixel camera ($1000 \times 1000 pix$) delivers nearly 300,000 independent (uncorrelated) 3D points in each camera frame.

A 3D sensor with these features allows for a continuous 3D measurement of moving or deforming objects, resulting in a *3D movie* (see e.g., tinyurl.com/3DMovCam12).[2] Like a hologram, each movie-frame encompasses the full 3D information about the object surface, and the observation perspective can be varied while watching the 3D movie (see e.g., tinyurl.com/3DMovCamView).[3]

The requisite low-cost technology is simple. The *single-shot ability*, paired with a *static pattern projection*, allows for the shape acquisition of *extremely fast live scenes*. Moreover, the sensor works very *efficient* from an information theoretical point of view. *Only two* properly positioned synchronized cameras are sufficient to solve the *profound ambiguity problem*, which is omnipresent in 3D metrology.

[2] Alternative link: tinyurl.com/3DCam-012.

[3] Alternative link: tinyurl.com/3DCam-view1.

Acknowledgements

It is my honest pleasure to thank the people that contributed to this work in so many different ways.

First and foremost, I express my gratitude to my doctoral advisor and mentor, *Prof. Gerd Häusler*. His curiosity and enthusiasm for the fundamental questions of nature have fascinated me since our first encounter in his lecture.

During the time I spent in his group, I was fortunate to learn so much from him. This does not only apply to physics, optics, and information theory, but also on how to engage with fellow colleagues or students—be it in lectures, personal talks, or other meetings.

I always enjoyed the (sometimes night-long) discussions with him. His talent for reducing complex issues to the essential important questions still impresses me. I consider myself very fortunate to have an advisor I can always trust, whose criticism I can learn and grow from, whose praise means an honest achievement, and whose company I have always enjoyed.

> *Lieber Gerd, als guter Lehrer und Freund hast du mich stets gefördert und gefordert. Dafür bin ich dir sehr dankbar!*

I thank *Prof. Gerd Leuchs* for accepting me into the Institute of Optics, and for the many possibilities I was provided to gaze over the horizon of my own area of research—be it in the numerous lectures or seminars at the Max Planck Institute for the Science of Light or at Ringberg Castle.

Moreover, I am grateful to *Ms. Margit Dollinger*, the secretary of the Institute of Optics, for the always competent and patient support in all things that a scientist does not so much like to do. Without her extensive expertise in organizational matters, I would have been lost in many situations.

The secretary of the OSMIN group, *Ms. Elizabeth Erhard*, I thank for the active support in all possible circumstances and for the introduction into the group-internal organizational processes. I always enjoyed our warm conversations.

I express my gratitude to the mechanic and electronic workshops of the University Erlangen-Nuremberg for their helpful cooperation and the great wealth of experience they offered and which I was able to take advantage of for the construction and electronic control of various prototypes.

A special thanks goes to my colleagues of the OSMIN group.

First and foremost, my '3D-cam-comrades' *Florian Schiffers*, *Yirui Jiang*, *Matthias Kuch*, and *Wei Wang* should be mentioned here. Without their commitment, the project in this form would not have been possible. In particular, I appreciate the help of *Florian Schiffers*, who evolved within the years from a 'Premium-HiWi' over a valued colleague to a good friend. His ubiquitous zest for action led to many important contributions to the project.

I am grateful to *Dr. Svenja Ettl*, *Dr. Oliver Arold*, and *Dr. Zheng Yang* for their excellent introduction to 'Flying Triangulation' at the beginning of my time at OSMIN, as well as for the stimulating technical discussions and wonderful experiences.

I also would like to thank the many undergraduate students, graduate students, interns, and assistants, I was allowed to advise and co-advise during my doctoral studies. In addition to the aforementioned gentlemen *Schiffers*, *Jiang*, *Kuch*, and *Wang* there are also *Dr. Franz Huber*, *Dominik Rauen*, *Max Schröter*, *Philip Dienstbier*, *Sarah Fouladi*, *Thomas Heid*, *Amirhossein Sadeghzadeh*, *Manuel Katanacho*, *Andreas Thurn*, *Benjamin Littau*, and *Max Perner*.

To all the other colleagues who accompanied me during my time at OSMIN—these are, above all, *Evelyn Olesch*, *Hanning Liang*, *Prof. Christian Faber*, *Dr. Ondrej Hybl*, *Markus Vogel*, *Roman Krobot*, *Christian Röttinger*, *Ralf Zuber*, *Dr. Alexander Bielke*, *Dr. Yuankun Liu* and *Dr. Markus Seraphim*—I express my graditude for the inspiring working atmosphere, the funny strolls to the cafeteria, our nice trips to conferences, and of course the enlightening discussions which always taught me new things.

A special thank you to *Prof. Christian Faber*, who contributed together with *Philip Dienstbier* significantly to the 'Tomografic Triangulation' experiment and with whom I enjoyed many technically stimulating and also very humorous experiences over the years.

I appreciate the second and the third reviewer of my thesis *Prof. Richard Kowarschik* and *Prof. Mitsuo Takeda* for the time they sacrificed, their valuable comments, and their highly accurate analysis.

I am grateful to *Mr. Raymond Hasler* for the language proofreading of this thesis. *Sänk jou Ray for pollisching my pour English!*

To my entire family and friends, I thank you for the great and endless support that you have given me.

In particular, I thank my parents who stand behind me in every decision. They have supported me throughout my entire education and have always encouraged me to be the best possible version of myself.

Finally, I express my deepest gratitude and admiration to my wife Anna for her understanding, trust, patience, and love.

Ich wüsste nicht was ich ohne dich machen würde! Dein Rückhalt, dein Mut und deine Liebe machen mich zu einem besseren Menschen!

Contents

Chapter 1
Introduction, Scope of Work and Summary of Results

The possibility to capture arbitrarily formed object surfaces in space in its full, *three-dimensional* shape, has become omnipresent. Although 3D measurement techniques have been used for many years in industry, medicine, forensics, or art, they have only recently found their way into public awareness—not least because of the introduction of *3D printers* and the steadily increasing computing power of *personal computers*. The latter facilitated the large scale marketing of *'virtual reality'* (VR) currently taking place.

Among the various 3D measurement techniques, the *optical* 3D metrology proves to be particularly advantageous. It captures the object surface contactless (= non-destructive) and is considered *'fast'*. Compared to a mechanical sampling of the surface, the word *'fast'* is certainly appropriate. However, the measurement time of available optical 3D sensors varies over several magnitudes. Consequently, *'fast'* is relative.

The ubiquity of digital 3D models in entertainment media and industry has also led to an increasing demand of potential costumers for high quality standards. Measurement results with (visible) deficits are no longer condoned. Buzzwords such as *'ultra-precise'*, *'high definition'*, *'high speed'* or *'large measurement field'* dominate the product sheets of many manufacturers. Frequently, a sensor is promised delivering all at once.

The physicist suspects that such a sensor is ***impossible***. Nature works with uncertainty relations. If one measure is maximized (e.g., high precision), other disadvantages (e.g., smaller measurement field, longer measurement time) have to be accepted. In other words:

> *'Nature gives no presents!'* [1]
> However, *one can 'buy something'*.

This thesis follows this guideline in a certain way. The goal is the development of a novel 3D measurement principle, which captures the shape of macroscopic

F. Willomitzer, *Single-Shot 3D Sensing Close to Physical Limits and Information Limits*, Springer Theses,
https://doi.org/10.1007/978-3-030-10904-2_1

objects with scattering, arbitrarily formed surfaces. The objects should be scanned ***fast, precise*** and ***with high resolution***—namely ***as good as physics and information theory allows***.

Since this task has a fundamental character, elaborate hardware components ('high-tech') such as high-speed cameras, precision optics, high-speed projectors, high precision translation stages or high-performance computers should be avoided as far as possible.

To still meet the required criteria, the following points must be considered:

Speed:
In order neither to be dependent of the frame rate of the camera in operation nor of the switching speed of a projector, a ***'single-shot' measurement principle***, paired with a ***static illumination*** has to be chosen. 'Single-shot' means ***no temporal sequence*** of camera exposures is used to generate a 3D height map.

Assuming a sufficiently bright light source, the time required for the acquisition of a full 3D model is solely dependent on the exposure time (not on the frame period) and can be kept very short. A static projection pattern favors the reduction of the exposure time. For instance a simple slide projector (projected pattern as a chrome-on-glass slide) can be used instead of a controllable DMD or FLCOS projector.

Precision:
In order to achieve a high precision of the measurement, the ***sources of the physical uncertainty in distance measurement*** have to be identified and minimized. The latter can be achieved, for example, by a reduction of the speckle noise and the application of a larger triangulation angle. This, however, has a negative effect on other parameters, e.g. the size of the measurement volume. A way must be found to cancel or extenuate the coupling of these parameters to a certain extent.

High Lateral Resolution and Data Density:
In order to resolve high frequent surface features and to acquire 3D data with high density, the object surface must be scanned ***independent of the neighborhood*** at ***as many points as possible***. For this purpose, a 'dense' pattern without spatial codification must be projected. However, the weak point of single-shot principles indeed is providing dense data. The achieved data densities are commonly no better than 1–3%. This problem stems from deep information theoretical reasons and can not be solved without additional information from other modalities. A new source of information has to be exploited to increase the data density.

To cope with the above mentioned tasks, the single-shot principle of *'multi-line triangulation'* is chosen as basis for the novel *'single-shot 3D movie camera'*. After the physical optimization of the sensor parameters, multi-line triangulation is able to acquire very precise data with a high lateral resolution along the projected lines. However, space is lost between the lines according to the profound ambiguity problem, which is ubiquitous in all 3D measurement principles. Consequently, also multi-line triangulation setups generally lack data density to a large extent, displaying values no better than $\sim 2\%$.

This thesis solves the ambiguity problem with the introduction of an additional modality: additional *perspective information*, provided by one or more *additional synchronized cameras*. The resulting approach preserves all valuable features of multi-line triangulation and simultaneously allows for the acquisition of 3D data with a high density. With nearly 30%, the sensor reaches values close to the theoretical limit. If 1-Megapixel cameras are applied, a 3D model consisting of 300,000 nearly uncorrelated 3D points is acquired during only one camera frame. This fact allows for the continuous 3D acquisition of moving or deforming objects, resulting in *'3D movies'* (see e.g. tinyurl.com/3DMovCam12).[1] Each movie-frame comprises the full 3D information about the object surface. The observation perspective can be varied while watching the 3D movie (see e.g. tinyurl.com/3DMovCamView).[2]

As proof of the principle, the novel method is implemented in a prototype setup. The measurement volume of the prototype has the dimensions $\Delta X \times \Delta Y \times \Delta Z = 300 \times 200 \times 100\ \text{mm}^3$ and is especially tailored to the acquisition of human faces. However, other objects with scattering surfaces in the same size can be measured as well. The *'single-shot 3D movie camera'* prototype acquires 3D models with a frame rate of 30 Hz and a precision better than $\delta z \leq 200\ \mu\text{m}$ in the entire measurement volume. From a detector with 1024×682 pixels, nearly 190,000 3D points are acquired during each camera frame.

Another basic aspect of this thesis is the *efficient* exploitation of the provided *channel capacity*. As a measure for the evaluation under information theoretical aspects, the so-called *'channel efficiency'* is introduced. Due to a special geometrical arrangement of the components and a novel evaluation method, the single-shot 3D movie camera requires only *two* synchronized cameras to provide the high data density. The related channel efficiency is much higher than the channel efficiency of other single-shot approaches. It is already comparable to the value for the *multi-shot approach* 'Phase Measuring Triangulation', which is considered as the 'gold standard' of optical 3D metrology.

The organization of this thesis is as follows:

Chapter 2 summarizes basic aspects of optics and information theory, which are necessary for the optimization and evaluation of the novel sensor principle.

Chapter 3 introduces the state of the art. The basic sensor principles commonly used to acquire macroscopic objects with scattering surfaces are discussed. Most of these principles rely on triangulation. The chapter pays special attention to how the correspondence problem is solved for each principle and it also refers to the three main desired specs: speed, precision and high resolution.

[1] Alternative link: tinyurl.com/3DCam-012, see also [2, 3].

[2] Alternative link: tinyurl.com/3DCam-view1, see also [2, 3].

Chapter 4 deals with line triangulation in particular. Using the concrete example of 'Flying Triangulation', considered of being the predecessor of the 'single-shot 3D movie camera', the ambiguity problem in line triangulation is explained.

Chapter 5 introduces a novel approach to solve the ambiguity problem in line triangulation by exploiting additional perspective information from one or more additional cameras. Several variations of the basic idea are discussed. Finally a prototype setup of the 'single-shot 3D movie camera' is presented.

Chapter 6 evaluates the novel sensor principle under physical and information theoretical aspects. In particular, key characteristics such as data density, precision, resolution and efficiency are discussed.

Chapter 7 introduces methods to further improve the measurement outcome. Besides the acquisition of color texture, a method is discussed how the data density can again be significantly increased by a simple geometrical trick.

Chapter 8 summarizes the algorithmic implementation of sensor calibration and 3D data evaluation. A modified calibration method, specially tailored to the 'single-shot 3D movie camera' is introduced.

Chapter 9 shows results, in particular 3D movies.

Chapter 10 collects comments, future prospects and continuative ideas.

The thesis concludes with *Chap.* 11.

References

1. G. Häusler, Fundamental limits of three-dimensional sensing (or: nature makes no presents). in *15th International Optics in Complex System, International Society for Optics and Photonics* (1990), pp. 352–353
2. Video file repository for Florian Willomitzer's Dissertation (urn:nbn:de:bvb:29-opus4-85442). University Erlangen-Nuremberg. nbn-resolving.de/urn:nbn:de:bvb:29-opus4-85442. Accessed 31 May 2017
3. Osmin3D, YouTube-channel of the OSMIN research group at the University Erlangen-Nuremberg. www.youtube.com/user/Osmin3D. Accessed 15 May 2017

Chapter 2
Selected Basics of Optics and Information Theory

The following chapter summarizes basics of optics and information theory. For the sake of compactness, only limits and relations are described, which play an important role in the physical and information theoretical optimization of the introduced sensor. Elaborate explanations and mathematical derivations are widely avoided. For a detailed explanation related literature is cited.

The sensor principle presented in this thesis is based on the well known principle of *line triangulation*. In the first section of this chapter, the basic idea of triangulation is explained. The next section introduces the physical limits of optical 3D sensors based on line triangulation. The third section specifies aspects for the information theoretical optimization of optical 3D sensors. The properties that make a 3D sensor *efficient* are discussed in particular.

2.1 Triangulation

This thesis intends to develop a new measurement principle for the fast three-dimensional acquisition of *macroscopic* objects with *scattering* surfaces. In addition to the commercially emerging 'time-of-flight'-techniques (to be discussed in Chap. 3), there is only one basic principle which is commonly used for this task: **Triangulation** (see Fig. 2.1): A *vertical* shift Δz of a surface point, observed with a camera under a defined angle - the *'triangulation angle'* θ - introduces a *lateral* shift $\Delta x'$ in the image plane on the camera chip.

Note that in this thesis all *image-sided* quantities are marked by primed characters, whereas *object-sided* quantities are not. Sometimes it is convenient to describe image-sided quantities in units of the pixel pitch. 1 pix is the distance between the centers of two neighbored pixels.

To identify the surface point in the camera picture, the object has to be sufficiently structured, e.g. by texture. If this cannot be ensured for all surfaces of interest, it is

F. Willomitzer, *Single-Shot 3D Sensing Close to Physical Limits and Information Limits*, Springer Theses,
https://doi.org/10.1007/978-3-030-10904-2_2

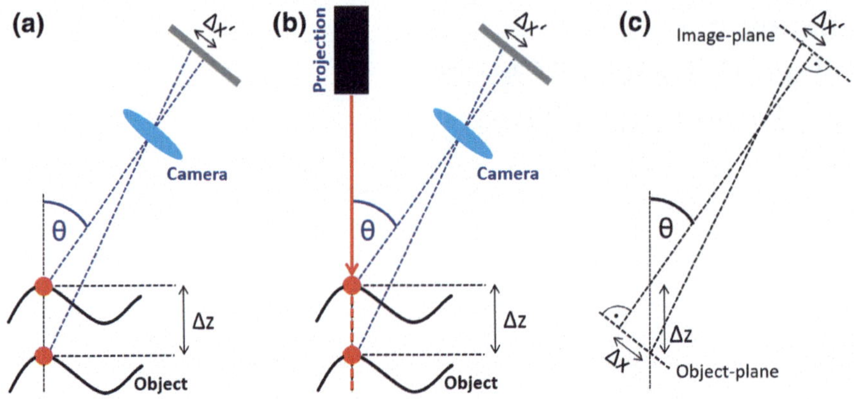

Fig. 2.1 Basic principle of triangulation: A vertical shift Δz of a surface point observed with a camera under the triangulation angle θ introduces a lateral shift on the camera chip. **a Passive** triangulation requires a sufficiently structured object surface. **b Active** triangulation structures the object artificially by proper illumination. **c** Sketch to the triangulation formula (Eq. 2.1). Δx is the related lateral shift in the object plane

possible to structure the surface artificially with the help of proper illumination, e.g. with a laser spot (see Fig. 2.1b).

For evaluation $\Delta x'$ is measured. Together with the known θ and magnification of the system β', Δz can be calculated via

$$\Delta z = \frac{\Delta x}{\sin\theta} = \frac{\Delta x'}{\beta' \sin\theta}, \tag{2.1}$$

where Δx is the related lateral shift in the object-plane of the camera.[1]

As an ultra-precise mounting of components would be required, the triangulation formula Eq. (2.1) is not used in practice without alteration. Moreover, each optical system suffers from aberrations, which are not considered in Eq. (2.1). Instead a calibration taking aberrations, varying magnification, and several other parameters into account is applied. The calibration for the measurement principle which is introduced in this thesis, is described in Chap. 8.

2.1.1 *Correspondence*

Correspondence describes the relation between two light patterns, e.g. the pattern to be projected and the pattern observed by the camera. This relation has to be found *unambiguously* in order to evaluate correct 3D data. The *'correspondence problem'*

[1]As can be seen in Fig. 2.1c, a vertical shift Δz introduces a shift of the object-plane as well. For sharp imaging over the whole range Δz, a certain *depth of field* is required. The shift of the object-plane also changes the magnification of the system ($\beta = \beta(z, \theta)$). These problems can be avoided by complying the *Scheimpflug condition*, which requires a tilt of the chip (= image plane) inside the camera [1–3].

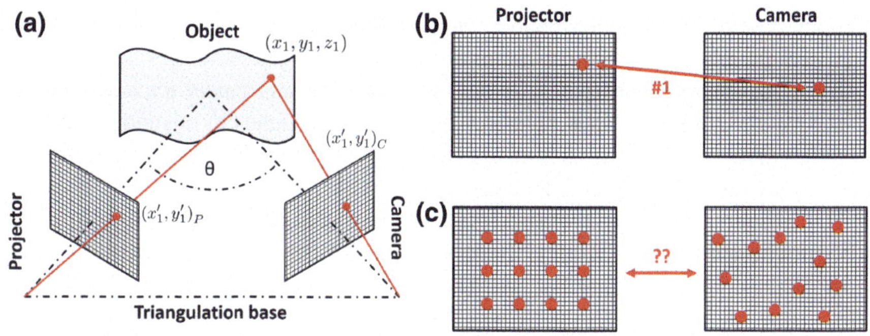

Fig. 2.2 Basic description of the correspondence problem

is well known in triangulation. Figure 2.2 introduces the correspondence problem. Point-shaped signals are assumed for simplicity but the explanation is also valid for stripes, lines, checkerboards or all other kinds of signal-patterns. A point projected from the coordinate $(x_1', y_1')_P$ in the projector plane, illuminates the object surface at (x_1, y_1, z_1), which is imaged by the camera at the position $(x_1', y_1')_C$ on the camera chip. If only one point is projected, it is obvious that the signal on the camera chip *'corresponds'* to this point (see Fig. 2.2a, b). The shift between $(x_1', y_1')_P$ and $(x_1', y_1')_C$ (often called 'disparity') can be calculated, which deciphers the depth z_1 in space. Generally this shift cannot be predicted or calculated in advance, since it is strongly dependent on the surface shape to be measured.

As soon as more than one signal (point) $(x_1', y_1')_P, \ldots, (x_n', y_n')_P$ is projected, the correspondence of points remains ambiguous (see Fig. 2.2c). This ambiguity cannot be resolved unless an ***additional source of information*** is exploited. Many options with different advantages and drawbacks are available, and indeed, one of the main distinctions between the multitude of triangulation principles is the *approach to solve the correspondence problem*. Representative methods are discussed in Chap. 3 and Sect. 10.1.

The measurement principle presented in this thesis is based on *line* triangulation: Instead of isolated points, *one or more lines* are projected onto the surface. Explanations about the principle and related problems are given in Chaps. 3 and 4. In Chap. 5 a new approach for the solution of the correspondence problem of line triangulation is presented.

2.2 Optical Basics of Triangulation Sensors, Condensed

2.2.1 Misleading Terminology: Measurement Uncertainty

All measurements include errors.

An *error* is defined by the difference between true value and measured value of a measurand. The exact value of an error is *always unknown*, otherwise it could be subtracted from the measured value. Only an *estimation* of the error can be given that is commonly called *'measurement uncertainty'*.

For a quantitative comparison of errors the term *'measurement uncertainty'* has to be properly defined. A lack of proper definition often leads to misunderstanding, since it remains unclear which kind of error and which statistical property (absolute value, standard deviation,...) is intended to be described.

The *Guide to the Expression of Uncertainty in Measurement (GUM)* [4] gives explicit rules for the terminology. Simplified, it can be said that two kinds of errors, showing different behavior exist: *systematic errors* and *random errors*. *Systematic errors* cause deviations from the true value, that are always the same for different repeated measurements. Therefore, they affect the so-called *'trueness'* of the measurement. In optical 3D metrology, a poor *trueness* can be caused by a poor calibration of the system, but by the object itself as well, e.g. if the light penetrates a certain distance into the surface before it is scattered (volume scattering). *Random errors* cause deviations from the true value that are different for different measurements. For a large number of measurements, the random errors tend to cancel each other. A large number of measurements can be achieved by repeated measurements of the same measurand or simultaneous measurements of different measurands with the same value. The standard deviation of a random error is defined as *'precision'*.[2] A poor precision of an optical 3D measurement is mainly caused by statistical noise on the data, e.g. speckle noise or photon noise (to be discussed in Sect. 2.2.4).

If the *trueness* and the *precision* are jointly considered, they define the *'accuracy'*, which describes how close a measurement is to the true value of a measurand. A common example for the graphical explanation of *trueness*, *precision*, and *accuracy* is shown in Fig. 2.3.

For the remainder of this thesis, the term ***measurement uncertainty*** δz is defined as the ***precision*** of 3D data.

One goal of this thesis is to achieve precise data by physical optimization of the sensor parameters. The *precision* of a single-shot line triangulation measurement

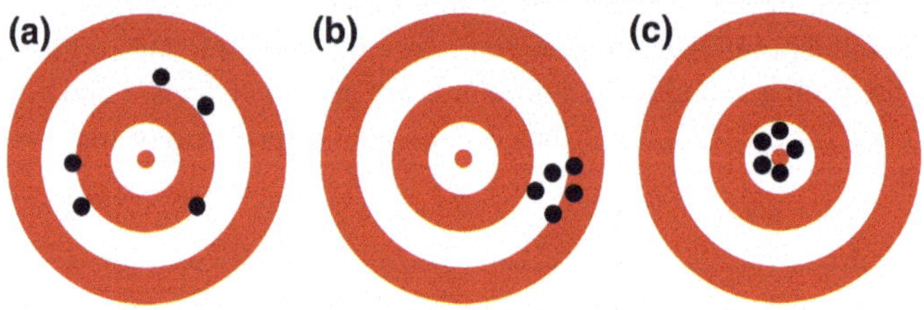

Fig. 2.3 Graphical explanation of trueness, precision and accuracy with target grouping. **a** Good trueness, poor precision, low accuracy. **b** Poor trueness, good precision, low accuracy. **c** Good trueness, good precision, high accuracy

[2]The exact definition is *'repeatability precision'*, if repeated measurements were performed under *constant* conditions and *'reproducibility precision'*, if repeated measurements were performed under *varying* conditions [5].

can easily be determined, e.g. via the large ensemble of points along a projected line (see Sect. 6.2).

An evaluation of the *trueness* or *accuracy* would require test objects with exactly known surface coordinates or comparative measurements from different sensors with very high accuracy. This is not the topic of this thesis, although the trueness of the calibration is briefly discussed in Chap. 8.

2.2.2 Misleading Terminology: Lateral Resolution

For an optical 3D sensor, several types of resolutions are important. Besides the *depth resolution*, which can be defined as the inverse *measurement uncertainty* $1/\delta z$, a 3D sensor displays *different kinds of lateral resolutions*. A *'lateral resolution'* is defined as $1/\delta x_r$, where δx_r is *'the smallest resolvable distance'*. Established definitions of this distance exist for 2D images: Depending on the quality of the objective, δx_r is limited by diffraction, lens aberration or the pixel pitch (see below). However, *'smallest resolvable distances'* exist in 3D images as well, where the definition of a lateral resolution is less intuitive and not as established as for 2D images. For a 3D image, δx_r can be defined as the simple lateral distance between two 3D points or the distance between two object features (small holes, etc.), that can be barely 'resolved' in the 3D data. Each definition of δx_r leads to a different kind of 'resolution'. In the following, these different possibilities are defined and discussed.

This is done for a certain reason: Data sheets of commercially available 3D sensors often display statements like *'Resolution:* 0.1 mm' (see e.g. [6]). Beside the usage of the inverse definition for resolution, such a statement leaves much room for interpretation. Is 0.1 mm the distance between each acquired 3D point or is 0.1 mm just the image-sided pixel size? Can this distance even be resolved by the objective?

This section aims to clarify these questions.

Optical Lateral Resolution

For an ordinary imaging system[3] the limit of the *optical lateral resolution* is given by diffraction. Rayleigh [8] defines the distance where two neighboring points can barely be resolved as the radius of the diffraction disc:

$$\delta x_{diff,R} = 1.22 \cdot \frac{\lambda}{2 \sin u_{obs}} . \tag{2.2}$$

Abbe defines the smallest resolvable line distance in a periodic pattern to

$$\delta x_{diff,A} = \frac{\lambda}{2 \sin u_{obs}} . \tag{2.3}$$

[3] Several methods (mostly in microscopy) for resolution beyond the diffraction limit exist. Probably the best known example is STED microscopy [7]. Such methods are not considered here.

In both equations, u_{obs} is the aperture angle of the imaging system and λ is the mean wavelength of the incoherent illumination. $\delta x_{diff,R}$ and $\delta x_{diff,A}$ vary only in the factor 1.22, which comes from the circularity of the diffraction disc.

For simplicity, the smallest optical resolvable distance δx_{diff} (analog for y-direction) is defined in this thesis by the Abbe-limit:

$$\delta x_{diff} = \delta x_{diff,A} = \frac{\lambda}{2 \sin u_{obs}} . \tag{2.4}$$

Note: A small δx_{diff} is *necessary*, but not *sufficient* for the reconstruction of high frequent signals. The *sufficient condition* for an artifact free signal reconstruction requires a *high sampling rate* as well, as expressed by the *Nyquist-Shannon sampling theorem* (to be discussed in Sect. 2.3.2).

Also note, that Eq. (2.4) yields only for the incoherent case and represents the absolute limit. Properly imaged patterns with the related cutoff-frequency $1/\delta x_{diff}$ already display zero contrast. Moreover, all objectives used in this thesis suffer aberrations and do not reach the diffraction limit. For images with aberrations, a criterion similar to the Rayleigh criterion can be applied [2], using the definition of the 'Strehl ratio' [9]. Further aspects of this topic are discussed in Sect. 2.3.2.

Lateral pixel resolution

An electronic camera does not know anything about the image quality of the objective lens. The camera chip samples the on-projected image with the lateral pixel pitch

$$\delta x'_{pix} = \frac{\Delta X'}{N_x} \quad ; \quad \delta y'_{pix} = \frac{\Delta Y'}{N_y} , \tag{2.5}$$

where $\Delta X', \Delta Y'$ is the chip-width (object sided: field-width) and N_x, N_y is the number of chip-pixels in x- and y- direction. If $\delta x'_{pix} > 2 \cdot \delta x'_{diff}$ (all considerations analog in y-direction), two neighboring pixels display widely *independent data*. This means that two pixels widely contain different information from different diffraction discs.

If an imaging system displays $\delta x_{pix} < \delta x_{diff}$, a further increase of the pixel resolution $1/\delta x_{pix}$ yields no advantage. It is even counter-productive since smaller pixels display a worse signal-to-noise ratio (to be discussed in the next section). This fact is often overseen in the design of consumer digital cameras or smartphone cameras - possibly on purpose, as a high pixel resolution (*'4K ultra high definition'*) is a good sales pitch.

So much for the resolution of *2D imaging*. The next sections are related to *3D imaging*. For the following definitions, it is assumed that each pixel displays 'independent' data and that the signals/objects are 'smooth enough' to not violate the sampling theorem.

Lateral Feature Resolution

As discussed in Sect. 2.1.1, the *correspondence problem* is crucial for nearly all triangulation principles. Many approaches solve the correspondence problem by shaping the projected light signals in a unique form, i.e. by projecting unambiguous *markers*. In the camera image, the marker is recognized by *correlation* with the related *'marker-function'*. Due to the spatial extent required for the correlation of the marker, only *one* independent 3D point can be calculated within the related area. This means that all three-dimensional object features (small holes, scratches,...) inside the correlation area *cannot be resolved* in the 3D data. The *lateral feature resolution* $1/\delta x_{feat}$, $1/\delta y_{feat}$ is reduced. For example, if markers of the size $5 \times 5\ pix$ are projected, the 3D data of object features, smaller than $5 \cdot \delta x_{pix} \times 5 \cdot \delta y_{pix}$ ('object-sided pixel distances') cannot be resolved.

Line triangulation is an interesting example for the discussion of feature resolution: For a vertical line, a widely *independent 3D point* is evaluated for each pixel-row of the line (within the limits of the sampling theorem), meaning that $1/\delta y_{feat} = 1/\delta y_{pix}$. To achieve better distance precision, the positions of the intensity maxima are evaluated in each row with sub pixel precision. This is done by an approximation with a Gauß-function, which needs additional space in x-direction (see explanation in Sect. 6.1). For a proper reconstruction, the line intensity profile on the camera chip has to display a width of at least $3 \cdot \delta x'_{pix}$, which is used for the approximation. This means, that the feature resolution in x-direction is no better than $1/\delta x_{feat} = 1/(\mathbf{3} \cdot \delta x_{pix})$.

Lateral Data Resolution

The sheer number of 3D points in a 3D height map does not indicate the quality of the sensor. Even if the sensor displays a low feature resolution, point clouds with a large number of 3D points can be generated by interpolation or similar methods. For the above example with the $5 \times 5\ pix$ correlation area, the application of a 'sliding window approach' would ensure the generation of a 3D point at each pixel.

Such approaches are common procedures to conceal poor data density, and *do not display independent 3D data*. It makes sense -for the sake of comparability- to consider *only independent 3D points* for the definition of the *lateral data resolution*. In that case $1/\delta x_{feat}$ defines an *upper limit* on $1/\delta x_{dat}$ (analog for y-direction): Independent data cannot be measured with a resolution better than the feature resolution.

Line triangulation is an interesting example for the data resolution as well, especially if more than one line is projected: Along the line direction each pixel displays a widely *independent 3D point*, resulting in $1/\delta y_{dat} = 1/\delta y_{pix}$. In x-direction the number of independent 3D points is only as large as the number of projected lines N_L. The resulting data resolution in x-direction is therefore

$$\frac{1}{\delta x_{dat}} = \frac{N_L}{N_x \cdot \delta x_{pix}} \,. \tag{2.6}$$

3D Data Density

For a quick and less abstract discussion of 3D point clouds, it is often sufficient to compare the number of independent 3D points N_{3D} in the point cloud and the number $N_{chip} = N_x \cdot N_y$ of camera pixels on the chip. In this thesis, this relation is defined as *3D data density*

$$\rho_{3D} = \frac{N_{3D}}{N_{chip}} . \tag{2.7}$$

The 3D data density can be interpreted as the 3D data resolution normalized with the pixel resolution.

$$\frac{\delta x_{pix} \cdot \delta y_{pix}}{\delta x_{dat} \cdot \delta y_{dat}} = \frac{N_{dat,x} \cdot N_{dat,y}}{N_x \cdot N_y} = \frac{N_{3D}}{N_{chip}} = \rho_{3D} \tag{2.8}$$

Thus, the 3D data density is *independent of the field width* and other scaling factors, and is well suited to *compare* different sensors (with different fields) with respect to their output of independent 3D data.

2.2.3 Surface Types

A multitude of sensor principles is available for optical 3D metrology. Altogether, these principles are capable of measuring nearly all kinds of objects. When the decision has to be made, which measurement principle is suited for a special measurement task, one property of the object to be measured, is of major importance and should be always examined first: the *micro structure* of the object surface which categorizes the *surface type*.

The surface type determines the interaction of the surface with the illumination, which might be reflection, coherent scattering or incoherent scattering. Together with all the different options for illumination (structured or homogeneous, monochromatic or colored, coherent or incoherent, directed or diffuse, polarized or unpolarized, temporally continuous or pulsed,...), a scope of several thousand variants for optical 3D sensors is possible.

The concepts presented in this thesis are based on triangulation. Triangulation only works on *scattering* object surfaces. For object surfaces with directed reflection, reflected light would only hit the camera for a proper surface normal, since the projected light only comes from approximately one point. The best scattering condition for triangulation sensors is *Lambert-scattering*. Here, the reflected *luminous intensity* I_r is described as $I_r(\alpha) = I_0 \cos(\alpha)$, where α is the observation-angle relative to the surface normal. This means, that the *reflected luminance*, given by $L_r = \frac{I}{cos(\alpha)}$ is *constant* for 'Lambert-scatterers' and can be described by

$$L_r = \frac{E \cdot \varrho}{\pi}, \tag{2.9}$$

with the *illuminance* E and the reflection coefficient ϱ [2]. In simple words: Lambert-scatterers appear equally bright from each direction. This characteristic is extremely favorable for triangulation principles. Probably this is the reason why plaster busts with white 'lambertian' surfaces are omnipresent in metrology- and computer vision publications, although they do not represent 'everyday objects'.

For opaque materials the surface scattering property is implicitly defined by the *surface roughness*. Volume scattering can be observed on partially translucent surfaces. Here the light is scattered from different layers under the surface. In that case the top surface layer need not necessarily to be rough in order to enable scattering.

A surface is optically *rough*, if the surface height variation δh within the area of one diffraction disc of the observation, is larger than $\frac{1}{4}$ of the illumination wavelength λ [10]. This means, that whether a surface is optically rough or not does not only depend on the surface itself. Wavelength and observation aperture[4] u_{obs} have an influence as well.

The limit-definition with $\frac{\lambda}{4}$ and the diffraction disc has a certain reason: For *smooth surfaces* ($\delta h < \frac{\lambda}{4}$) all reflected elementary light waves within the diffraction disc will be superimposed during image formation *solely* by constructive interference. The result is a well-defined point spread function in the image plane. For *rough surfaces* ($\delta h > \frac{\lambda}{4}$) the reflected elementary waves display large random phase differences. For coherent illumination this results in *speckle* in the image plain.

2.2.4 Speckle - the Ultimate Enemy of Metrology?

Speckle can be observed when a *rough* surface is illuminated with (partially) *coherent* light. The granular structure of a speckle field is *completely random* and can not be predicted for an unknown surface to be measured. Details about speckle formation are given in [10–16] and many others. For metrology *subjective speckle* play the more important role, since an imaging system is involved in nearly all applications.

A few methods exploit speckle. Examples are 'phase shifting speckle interferometry' or 'speckle photography' [17]. For most other methods speckle is an 'enemy', since *speckle significantly spoils the image quality*. In many cases speckle becomes the 'ultimate enemy'. One of these cases is line triangulation. In [18], it is shown that for coherent illumination *speckle noise* can exceed all other noise sources, i.a. *quantization noise*, *read out noise*, *PNRU (Photon Response Non Uniformly) noise* or *shot noise*. This is even the case if white light is used for the illumination (which is often misinterpreted[5] as 'incoherent').

[4]For example a ground glass observed with a microscope will appear like many tiny mirrors. Observed with the naked eye, it appears mat.

[5]For the surface structures to be measured in this thesis, white light might lead to a low *temporal coherence* (see also Eq. (2.16)). The *spatial coherence* is mainly dependent on the used apertures (see also Eq. (2.18)).

If speckle noise is the dominant noise source, it puts an ultimate limit on the measurement uncertainty. This can be explained as follows: Due to [10, 16][6] the speckle noise induced *lateral uncertainty* $\delta x_{speckle}$ for the localization of the image of a light spot on a rough surface can be described in object space as

$$\delta x_{speckle} = \frac{C_s}{2\pi} \cdot \frac{\lambda}{\sin u_{obs}} , \tag{2.10}$$

Obviously, $\delta x_{speckle}$ is proportional to the object sided spot diameter of Eq. (2.4). The proportionality factor includes the *speckle contrast* C_s, which is the inverse signal to noise ratio (*SNR*)

$$C_s = \frac{1}{SNR} = \frac{\sigma_I}{\bar{I}} . \tag{2.11}$$

$\bar{I}$ is the mean intensity of the signal, σ_I is the corresponding standard deviation.

Via triangulation (see Eq.(2.1)) of the line signal $\delta x_{speckle}$ leads finally to the *absolute limit of distance measurement uncertainty*:

$$\delta z_{speckle} = \frac{C_s}{2\pi} \cdot \frac{\lambda}{\sin u_{obs} \sin\theta} \tag{2.12}$$

In the worst case, signal and noise have the same level, resulting in $C_s = 1$. This is the case for many commercially available line triangulation scanners. Surprisingly, features such as *'laser illumination'*, *'small apertures'* (large depth of field) or *'ultra high chip resolution'* are advertised. As the following discussions reveal, these features give rise to higher speckle noise and should better not appear in an advertisement.

For the discussion of speckle-induced noise, the speckle contrast C_s has to be reviewed in more detail. The results of this review will lead to a *'recipe' for the physical optimization of line triangulation scanners*, given in the next section.

The speckle contrast is a product of four factors: The spatial factor c_{spat}, the temporal factor c_{temp}, the polarization factor c_{pol} and the pixel factor c_{pix} [19].

$$C_s = c_{spat} \cdot c_{temp} \cdot c_{pol} \cdot c_{pix} \tag{2.13}$$

The *polarization factor* c_{pol} describes the influence of polarized light on the speckle contrast. As shown in [20], the speckle contrast is reduced by a factor of $1/\sqrt{2}$ at completely depolarizing surfaces. An additional unpolarized illumination leads to an additional contrast reduction by the factor of $1/\sqrt{2}$. This means that

[6] Apart from the theoretical derivation from speckle statistics, the authors of [16] show that Eq. (2.10) can be also derived from Heisenberg's uncertainty principle. The localization uncertainty of a laser spot at a rough surface ($1/SNR = C_s = 1$) is surprisingly the same as the localization uncertainty of a single photon without influence from a scattering surface ($1/SNR = \sqrt{n} = 1$).

$$c_{pol} \leq \frac{1}{2}\,, \tag{2.14}$$

whereat the maximal reduction $c_{pol} = 1/2$ is reached for complete depolarization.

The pixel factor c_{pix} describes the influence of the pixel size on the speckle contrast. As long as the pixel width d'_{pix} of the camera chip in use is *smaller* than the diameter of an image sided subjective speckle, c_{pix} equals 1. For pixels *larger* than the size of a subjective speckle the intensity variation of more than one speckle is integrated over one pixel, which results in a contrast reduction. Due to [11] the size of a subjective speckle is approximately the size of a diffraction disc (Eq. (2.2)) on the chip.[7] c_{pix} can therefore be approximated [21] as

$$c_{pix} = \min\left(\frac{\lambda}{d'_{pix} \cdot \sin u'_{obs}},\ 1\right). \tag{2.15}$$

In most cases a reduction of c_{pix} is only of limited use, since it is accompanied by a reduction of the lateral resolution.

The remaining factors c_{temp} and c_{spat} have the largest potential to reduce C_s. During the process of 'physical optimization' of a sensor (see next section), particular attention has to be paid on these two factors.

The *temporal factor* c_{temp} describes the influence of the *temporal coherence* on the speckle contrast and is expressed as

$$c_{temp} = \left(1 + \left(\frac{4\,\delta h}{l_c}\right)^2\right)^{-1/4}. \tag{2.16}$$

It was already mentioned that c_{temp} can be reduced by a light source with low temporal coherence. A significant reduction of c_{temp} is achieved if the coherence length l_c is smaller than the surface height variation δh within one diffraction disc. Note that δh equals the surface roughness σ_h (standard deviation of the surface height) only for opaque untilted surfaces. For surfaces with a significant tilt, the height variation within a diffraction disc is larger than the surface roughness σ_h. In triangulation, where the optical axes of illumination and observation enclose the triangulation angle θ, this is possible, too. As explained in [19], δh and can then be expressed as $\delta h = 2\,\delta x_{diff,R} \cdot \sin\theta$. For surfaces with volume scattering properties, δh represents the penetration depth.

The *spatial factor* c_{spat} describes the influence of the *spatial coherence* on the speckle contrast. If an object surface is illuminated by a light source with half diameter d_{ill} from the distance z, the width of the coherence function on the object surface is given by

[7] The definition for the diameter of a single subjective speckle varies in literature. For example, the author of [14] calculates $d'_{s,subj}$ to *half* of the diffraction disc diameter.

$$d_\gamma = 1.22 \cdot \frac{\lambda z}{d_{ill}} \approx 1.22 \cdot \frac{\lambda}{\sin u_{ill}} . \tag{2.17}$$

In an actual projection system, d_{ill} is the half diameter of the aperture of the projection lens. If d_γ is the same size or larger than the object-sided diffraction disc $2\ \delta x_{diff,R}$ of the observation (see Eq. (2.2)), all elementary waves inside the diffraction disc are fully coherent, which results in $c_{spat} = 1$. For $d_\gamma < 2\ \delta x_{diff,R}$ elementary waves inside the diffraction disc are superimposed incoherently, which results in a reduction of c_{spat} by $c_{spat} = d_\gamma/(2\ \delta x_{diff,R}) = \sin u_{obs}/\sin u_{ill}$. Therefore, c_{spat} can be expressed as

$$c_{spat} = \min\left(\frac{\sin u_{obs}}{\sin u_{ill}},\ 1\right) . \tag{2.18}$$

In order to achieve a small *absolute limit of distance measurement uncertainty* $\delta z_{speckle}$, the *lateral uncertainty* $\delta x_{speckle}$ and thus C_s has to be decreased. The next section summarizes all possibilities for a reduction of $\delta z_{speckle}$.

2.2.5 *Physical Optimization*

The relations discussed in the last section display several possibilities for the reduction of coherent speckle noise, which is the dominant noise source for many triangulation principles. In the following, a ***'recipe' for the physical optimization*** of such principles is given. The 'ingredients' for this 'recipe' are found in the equations of the last section.

- ***Short coherence length!***
  The influence of temporal coherence is given in Eq. (2.16). As the user has no influence on the roughness or volume scattering properties of the measured surface, the coherence length of the used light source has to be chosen as small as possible. In practice, this results in the *utilization of broad band light sources*, such as white light LEDs. No monochromatic lasers! This is particularly effective for surfaces with higher surface roughness or volume scattering.
- ***Unpolarized illumination!***
  In order to reach a maximal reduction of the polarization factor, the projector has to illuminate the object surface with unpolarized light. Polarized lasers or LCD projectors are unfavorable. Many so-called 'LED-projectors' use an LCD panel for image generation. An LED is only used to provide background illumination. Hence, such projectors are *polarized* light sources.
- ***Sufficiently large pixels!***
  As shown in Eq. (2.15), a pixel size larger than the size of a subjective speckle leads to a contrast reduction. On the other hand, the size of the subjective speckle is equal to the size of the diffraction disc. This means that a contrast reduction would sacrifice lateral resolution. In practice, a trade-off between contrast reduction and lateral resolution has to be found.

- ***Large apertures!***
  As shown in Eq. (2.18), the illumination aperture has to be *larger* than the observation aperture in order to reduce the spatial speckle contrast. According to Eq. (2.12) and Eq. (2.15), the observation aperture should also be large.
  If, by proper choice of the parameters, a reduction of c_{spat} and c_{pix} is assumed, the fundamental localization uncertainty $\delta x_{speckle}$ (Eq. (2.10)) can be rewritten as

$$\delta x_{speckle} = \frac{c_{temp}\, c_{pol}}{2\pi \cdot d_{pix}} \cdot \frac{\lambda^2}{\sin u_{ill}\, \sin u_{obs}} . \tag{2.19}$$

  This results in the condition *'all apertures as large as possible, while* $\sin u_{ill} >$ $\sin u_{obs}$*'*. Again, no lasers! ($\sin u_{ill} << \sin u_{obs}$).

If an optical system is adjusted as discussed above, the speckle noise is decreased significantly and might not be the dominant noise source anymore. The contribution of other noise sources, for instance *electronic noise* or *photon noise*, becomes more important. According to [18], the new dominant noise source is the *photon noise*, also known as *'shot noise'*. The noise level of related sensors is then called *'shot noise-limited'*. The physical property of the photon noise defines the next 'ingredient' of the 'recipe':

- ***Many photons!***
  For shot noise limited systems the *SNR* is defined by the statistical value $SNR = 1/\sqrt{n}$, with n being the number of incoming photons (detection efficiency of ~ 1 assumed). In practice, a large *SNR* is achieved by *bright light sources*, *large pixels* or a *long exposure time*, whereas the first possibility is the most favorable.

After the optimizations described above, the resulting overall localization uncertainty δx of the light signal is *significantly smaller* than $\delta x_{speckle}(C_s = 1)$ (fully coherent case). As 3D metrology deals with *distance measurement*, the *localization uncertainty* δx finally has to be transformed into a *distance uncertainty* δz. This is done via triangulation with the triangulation angle, which is the last 'ingredient' of our 'recipe' for a small distance uncertainty δz.

- ***Large triangulation angle!***
  This optimization parameter differs from the already discussed parameters, as it has no influence on the lateral localization uncertainty δx. Due to [16] or Eq. (2.1), the distance uncertainty δz of a triangulation sensor is calculated via

$$\delta z = \frac{\delta x}{\sin \theta} . \tag{2.20}$$

  The resulting (trivial) perception is very important for the discussions in the following chapters: Two equal triangulation sensors which *only differ in their triangulation angle* would display *the same* δx (or $\delta x'$ on the camera chip). Nevertheless, the sensor with the *larger triangulation angle* would display a *smaller distance uncertainty* δz. Therefore, the *triangulation angle* θ *should be chosen as large as possible*.[8]

The power of the given optimization-recipe, especially for the reduction of speckle noise, is demonstrated in Fig. 2.4. The 3D data of a single line profile is measured with two different sensor setups (a) and (b). The measured object is a white, planar surface with only little volume scattering. For the sake of comparability, the image sided observation aperture is chosen approximately the same in both measurements, which would result in the same subjective speckle size for fully coherent illumination. The exposure limit is also the same. The sensor setups exploit different *illumination*. Setup (a) employs laser illumination with a large coherence length and $\sin u_{ill} << \sin u_{obs}$. Setup (b) employs a white light LED projector with a large aperture. For setup (a) as well as for setup (b) the left column shows an imaged line segment of ~75 pix length. The middle column displays the noise in x-direction, which defines the lateral localization uncertainty δx. For each pixel-row, the sub-pixel precise position of the line maximum is evaluated via Gauß-correlation and plotted in red on top of the

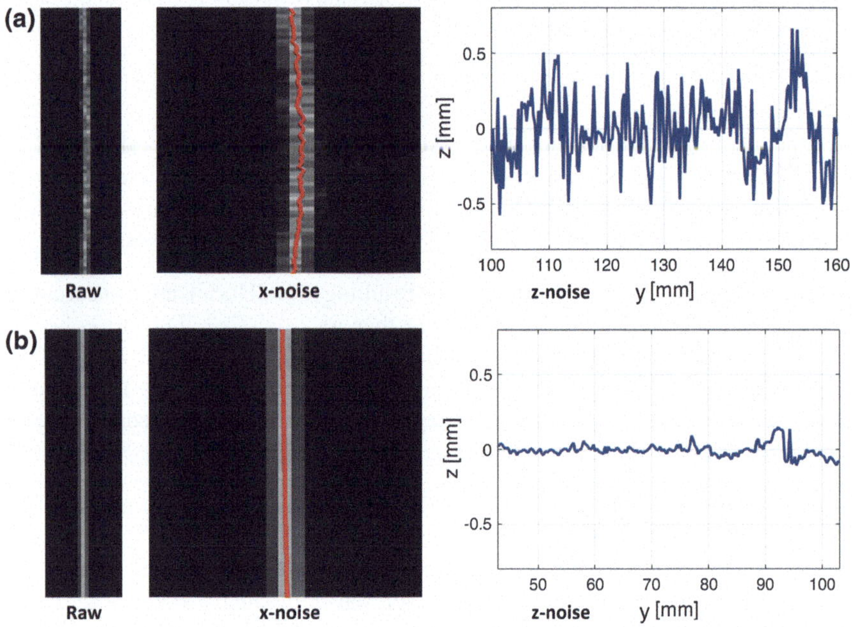

Fig. 2.4 Physical optimization for line triangulation setups. **a** Coherent illumination with laser. **b** Reduced coherence. (**a**) and (**b**): Left: imaged line segment of ~ 75 pix length. Middle: imaged line segment stretched in x-direction. Evaluated sub-pixel precise line maximum plotted on top. Right: Related distance values, evaluated with an assumed triangulation angle of $\theta = 9°$ and a field width of $\Delta X = 300$ mm

[8]Unfortunately this can not always be realized, as a large triangulation angle increases occlusions, makes the device less compact, and -as will be shown in Chap. 4- can reduce the measurement depth. In practice a tradeoff has to be found.

raw-data (stretched x-scale for better visualization). For the calculation of the z-noise via the triangulation formula (right column, stretched z-scale for better visualization), a triangulation angle of $\theta = 9°$ and a field width of $\Delta X = 300$ mm is assumed. As expected, the *coherent setup* (a) displays a *significant higher noise level*, which is in accordance with the given 'optimization-recipe'.

As already mentioned, some of the ingredients of this 'recipe' may result in certain tradeoffs. A good example is the specification of the aperture angles u_{ill} and u_{obs}: In practice a certain depth of field is required. This puts an upper limit on the aperture sizes. Other examples for technical tradeoffs may be triangulation angle versus occlusion or device compactness, as well as pixel size versus lateral resolution.

What is physical optimization?

An optical 3D sensor is ***physically optimized****, if all of the mentioned* ***ingredients are taken into account*** *and if the resulting* ***tradeoffs are well-defined and tailored to the specific application****. Then, the sensor acquires data from the object surface* ***as good as physics allows****.*

2.3 A Few Information Theoretical Aspects of Line Triangulation Sensors

Information theory developed by Shannon [22, 23], was originally designed as a tool to investigate fundamental limits in signal processing and communication. Meanwhile information theory is also used in many other disciplines.

As already shown in several publications, i.e. [21, 24, 25], Shannon's information theory is a *powerful tool* to *describe and optimize optical 3D sensors*.

In the following, some basics of information theory necessary for the optimization of line triangulation sensors under information theoretical aspects are discussed.

2.3.1 *Transmission Model of Optical 3D Sensors*

In information theory the transmission of a signal is described by the transmission model for communication channels [22]. As shown in Fig. 2.5a, a communication system can be described in general by *source*, *transmitter*, *channel*, *receiver* and *drain*. The *transmitter* consists of the *source encoder* and the *channel encoder*, whereas the *receiver* performs the inverse operations by its components *channel decoder* and *source decoder*. The information is transferred from the transmitter to the receiver via the *noise affected channel*. A-priori information helps to synchronize transmitter and receiver.

As discussed, *optical 3D sensors* can be described by this transmission model, too [21, 24, 25]. The related analogy is shown in Fig. 2.5b. Here, the *object surface* plays

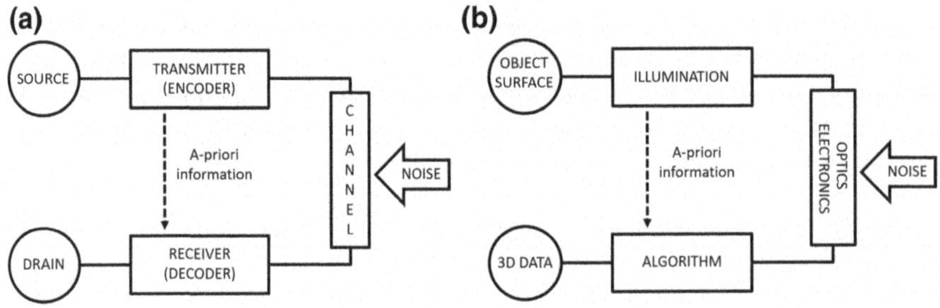

Fig. 2.5 Information theoretical transmission model. **a** General model for a communication system described by Shannon [22]. **b** Shannon's transmission model applied to optical 3D sensors

the role of the information source. The surface shape is encoded into transferable signals by proper *illumination*. Later it will become clear that *proper source encoding* is a very crucial operation in 3D surface measurement. In line triangulation the (unknown and not transferable) object surface shape is encoded in *positions of points along deformed lines*, which can be transferred through the *noise affected channel* consisting of *optics and electronics*. With an *algorithm* the encoded object surface is decoded into *3D data*. An example for *a-priori information* is the specification, that the projected *lines are straight*. This information is *not* transferred trough the channel and known by the algorithm in advance. It ensures, that a *deformation* of an imaged line is interpreted as a *variation* of the surface shape.

If the class of objects to be measured (and thus their estimated shape) is known in advance, a-priori information can be exploited to a large extent by source encoding to improve the outcome of the measurement. Of course, then the *flexibility* of such sensors is more restricted.

2.3.2 *Bandwidth Restrictions and Sampling Theorem*

The *Nyquist-Shannon sampling theorem* plays an important role for high frequency signals being reconstructed after imaging on a chip with discrete pixels. In simplified terms,[9] it can be said that a signal with the maximal frequency ν_{signal} can be reconstructed without artifacts, if the (pixel) sampling frequency ν_{sample} is

$$\nu_{sample} \geq 2 \cdot \nu_{signal} \ . \tag{2.21}$$

In practice ν_{sample} is a fixed value of the imaging system defined by the pixel pitch. Then,

$$\nu_{nyq} = \frac{\nu_{sample}}{2} \tag{2.22}$$

[9]Baseband signal ($\nu_0 = 0$) as well as infinite large fields and object surfaces assumed. For detailed conditions and derivations review [23].

is the *highest frequency component* of the signal that can be reconstructed after imaging - the so-called *'Nyquist-frequency'*. If the sampling theorem is violated (*'undersampling'*), the signal reconstruction yields artifacts. This so-called *'aliasing'* arises, because undersampling creates new frequency components by mixing of low and high frequencies.[10]

Fortunately, undersampling can be widely avoided, even if the surface (and therefore the highest frequencies) is not known in advance. In signal processing operations this is done by a low pass filter ('anti-aliasing filter'). The filter is applied *before* the signal is sampled. Ideally only frequencies $\leq \nu_{nyq}$ pass.

In imaging the lens can act as a low pass filter. By slightly defocusing the lens, its lateral resolution disc with radius δx_{lens} increases, meaning that virtually only frequencies $\nu_{lens} \leq 1/\ \delta x_{lens}$ pass. To avoid undersampling, ν_{lens} has to be smaller than the Nyquist-frequency. However, defocusing is commonly only necessary for diffraction limited lenses and image sensors with large pixels. Mostly the aberrations of the lens already ensure $\nu_{lens} < \nu_{nyq}$.

Where is the limit? – In other words: What is the maximal object frequency bandwidth $B = \nu_{max} - \nu_0$ that is delivered by the 'optical channel' (lens plus imager)?

As shown in Eq. (2.4), δx_{diff} defines the smallest resolvable distance for imaging systems. The maximal required sampling frequency in x-direction is therefore

$$\nu_{sample,x} = \frac{N_x}{\Delta X} \stackrel{!}{=} \frac{2}{\delta x_{diff}}, \tag{2.23}$$

where N_x is the number of pixels and ΔX is the field width in x-direction (analog for y-direction). For a circular lens, the area of frequencies to be transferred is circular in the Fourier domain. Considering Eq. (2.22) the *maximal transferable space-bandwidth product* can be written as

$$(B_x \cdot B_y)_{max} = \frac{\pi}{4} \cdot \frac{\nu_{sample,x}}{2} \cdot \frac{\nu_{sample,y}}{2} = \frac{\pi}{\lambda^2} \cdot \sin^2 u_{obs}. \tag{2.24}$$

This product plays a crucial role for line triangulation (and other metrology principles as well): The *shape of the object* is encoded in *positions of lines*, which are projected onto the object surface. If these lines are projected onto steps or sharp edges of the surface, the 'item' to be imaged (surface with on-projected lines) contains very high frequency components. According to Eq. (2.24) these frequencies *cannot be transferred* through the optical channel and will not be present in the evaluated 3D model. This fact should always be kept in mind when talking about 3D measurements. *Sharp edges can never be reconstructed without artifacts!*

This is especially the case when the continuity of the line signal is interrupted ('jump edges'). The low-pass filtering effectuates that a line signal is also present shortly beyond the edge, resulting in ambiguous 3D data at the effected region. 'Roof

[10] Well known examples for aliasing phenomena are the flickering of striped shirts on TV or an apparent 'waviness' of brick walls in digital images.

edges' do not interrupt the continuity of the line signal. In that case a sharp edge becomes only rounded in the 3D data as a result of the low-pass filtering (see Sect. 6.3 for further discussions).

Note that Eq. (2.24) displays the absolute limit of the space-bandwidth product. In presence of lens aberrations the product is smaller. Also note that, according to Sect. 2.2.2 the evaluation of 3D data might involve correlation in the 2D images, which further decreases the maximal frequency components in the 3D image.

Besides the 'optical channel' the encoded information must pass the 'electronic channel' (signal transportation) and the 'discrete channel' (signal digitization) as well. These channels also introduce bandwidth restrictions. The resulting space-time-bandwidth product is the minimal bandwidth product of *all involved channels*. This is especially relevant for the temporal bandwidth. The immensely high value of the optical channel given by

$$(B_t)_{max} \approx \frac{c}{l_c} \approx 10^{14}\ Hz \qquad \text{(for white light)}\ , \tag{2.25}$$

with the speed of light c and the coherence length l_c [21], can never be transferred trough the other channels.

In the next section a measure to judge the performance of the complete channel is introduced–the *'channel capacity'*. In the second part of that section, the concept of the channel capacity is used to define a quantitative measure for the performance of optical 3D sensors under information theoretical aspects – the *'channel efficiency'*.

2.3.3 Channel Capacity and Channel Efficiency

The *channel capacity* limits the *highest rate of information*, that can be *reliably transferred* over a certain AWGN (*'additive white Gaussian noise'*) communication channel. Commonly, the channel capacity is measured in *bit/s*. However, for the purpose of this thesis it is more useful to discuss the *total amount of transferred information*, which is given by the time-normalized channel capacity (in *bit/s*) multiplied with the acquisition time ΔT. In the following, the term *'channel capacity'* Γ will be used to express exactly this *total amount of transferred information*.

According to [21] the channel transmission model of an optical sensor can be described by a channel with discrete values in space and time [26]

$$\Gamma = \frac{1}{2}\, N_x\, N_y\, N_t\, \log_2(1 + SNR)\ , \tag{2.26}$$

with N_x, N_y, and N_t being the sampling steps in x- and y- direction and in time t. *SNR* is the signal to noise ratio of the input signal.

Due to the sampling theorem (Eq. (2.22)), the maximal transferable bandwidth in x-direction (analog for y- and t- domain) is

$$B_x = \frac{\nu_{sample,x}}{2} = \frac{N_x}{2\,\Delta X}\,. \tag{2.27}$$

Consequently this leads to

$$\Gamma = 4\,\Delta X\;B_x\;\Delta Y\;B_y\;\Delta T\;B_t\;\log_2(1+SNR) = 4\cdot STBP\cdot\log_2(1+SNR)\,. \tag{2.28}$$

$STBP$ is the *'space-time-bandwidth-product'*. For the described example of the optical channel, the channel capacity is therefore

$$\Gamma_{opt} = \frac{4\pi\sin^2 u_{obs}}{\lambda^2}\;\Delta X\;\Delta Y\;\cdot\Delta T\;B_t\;\log_2(1+SNR)\,. \tag{2.29}$$

Obviously, the 'recipe' for a *good optical sensor* given in Sect. 2.2.5, can also be found in the equation above: *large aperture, large detector area* and a *large SNR*. This means a physically optimized sensor is also able to transfer much information – at least through the optical channel.

The channel capacity of the complete system is always smaller as the minimum of the optical, electrical, and discrete channel capacity. The channel capacities of the electrical and discrete channel are discussed in [21, 24, 25]. The channel capacity of the complete channel can still be described by Eq. (2.26), where the argument of the logarithm includes the added noise of all sub-channels [21].

How can the concept of the channel capacity be exploited to evaluate the potential of optical 3D sensors under information theoretical aspects?

Each sensor acquires one or more 2D images for the evaluation of one '3D image'. The *maximal information content* $\mathcal{M}_{2D}$ of these 2D images can be described by Eq. (2.26).

$$\mathcal{M}_{2D} = \frac{1}{2}\;N_x\;N_y\;N_t\;\log_2(1+SNR) \tag{2.30}$$

The variables in this equation can be *assigned to 3D images* as well! In analogy to the *maximal information content* $\mathcal{M}_{2D}$ for one or several 2D images, the right side of Eq. (2.30) can also be used to determine the *maximal information content* $\mathcal{M}_{3D}$, *that can be comprised by a 3D height map.*

For the intended evaluation of an optical 3D sensor under information theoretical aspects, it is of great interest *how much 3D information can be maximally obtained from the provided 2D dataset*. Due to the different *SNR* for 2D and 3D images, a comparison of the sheer number of 2D pixels and 3D points (= data density ρ_{3D}) is too simple in this case. A much better measure is the *comparison of the maximal information contents*. The resulting ***channel efficiency*** η is defined as:

$$\eta = \frac{\mathcal{M}_{3D}}{\mathcal{M}_{2D}} \tag{2.31}$$

In $\mathcal{M}_{2D}$, the factor $N_x \cdot N_y$ describes the total number of provided pixels and $N_t = F$ is the number of successively acquired camera frames required for *one* 3D height map. If only one camera is used for image acquisition, $N_x \cdot N_y$ is the number of pixels on the chip. For simultaneous acquisition with more cameras yields $N_x \cdot N_y = Q \cdot N_{chip}$, where Q is the number of simultaneously acquired images (cameras of the same type/pixel resolution assumed).

In $\mathcal{M}_{3D}$, $N_x \cdot N_y$ is the number of acquired independent 3D points N_{3D}. Of course, only *independent 3D points* must be considered. Whether a 3D point is independent or not is mainly determined by the *feature resolution* of the sensor. Interpolated data or data from a common correlation window are *not independent*!

The channel efficiency of Eq. (2.31) is *not* dependent on the number of successively acquired 3D height maps. Therefore, only *one* height map is considered ($N_t = 1$ in $\mathcal{M}_{3D}$). The *SNR* of the 3D height map is given by the *dynamic range* which is defined by the ratio of measurement depth ΔZ (range of 3D data) and the noise δz of the data. Consequently η can be expressed as:

$$\eta = \frac{\mathcal{M}_{3D}}{\mathcal{M}_{2D}} = \frac{\rho_{3D}}{Q\,F} \frac{\log_2(1 + \frac{\Delta Z}{\delta z})}{\log_2(1 + SNR)} \tag{2.32}$$

Commonly, the dynamic range of 3D height maps is larger than the *SNR* of a 2D image. Nevertheless, $\mathcal{M}_{3D}$ is smaller than $\mathcal{M}_{2D}$. This has deep information theoretical reasons: Each 2D image comprises information that is *not used* for the calculation of 3D data by triangulation. Besides the (wanted) *position information* (e.g. position of the line intensity maximum on the chip), each image contains as well information about the *reflectivity of the object surface* and the *background illumination*. This information contributes to $\mathcal{M}_{2D}$, but not to $\mathcal{M}_{3D}$. Hence, it is expected that the channel efficiency is $\eta \leq 1/3$ for all measurement principles, using *only the position information* of a signal for 3D data generation. This is the case for most triangulation principles.

An example: A sensor with two 1-Megapixel cameras (8 bit) acquires a point cloud of 100,000 independent 3D points within a measurement depth of 100 mm. The sensor needs 3 subsequent images from each camera. For each camera image the $\log_2(1 + SNR)$ is around 6 bit. The noise of the 3D data is 100 µm.

According to the specifications above the sensor would be estimated quite good. It displays good precision and data density and it needs only three subsequent images for a 3D point cloud. Nevertheless its channel efficiency is only

$$\eta = \frac{0.1}{2 \cdot 3} \frac{\log_2(1 + 100/0.1)}{6\,bit} = 2.8\% \,. \tag{2.33}$$

In Sect. 6.5, the channel efficiency of the sensor presented in this thesis will be calculated and compared with the channel efficiency of other sensor principles using triangulation. The comparison will reveal that many state-of-the-art 3D sensors have a channel efficiency not better than ~1–3%.

Remark 1: As mentioned above, the formula for the channel capacity was derived by Shannon under the assumption of *additive white Gaussian noise*. Speckle noise on the other hand is *multiplicative* in most cases [27]. However, the noise properties of a 3D image and the related 2D image(s) are significantly different: The 3D noise (depth error) that is introduced by speckle noise in triangulation systems (see Eq. (2.12) and [10, 16]) is essentially additive, meaning that the dynamic range $\Delta Z/\delta z$ (which is quantized with the noise intervals) becomes linearly spaced for most sensors since the working distance is much larger than the measurement depth ΔZ. For the maximal information content $\mathcal{M}_{2D}$ of a 2D image taken with an 8 bit camera, the value $\log_2(1 + SNR)$ is set to 6 bit for the remainder of this thesis. As discussed, the SNR is significantly better than unity, as the sensor is optimized for low coherence. The value of 6 bit is considered as a 'good estimate' and is in accordance with empirical findings from a variety of experiments performed at the Erlangen Institute of Optics. It can be also found in good approximation for the sensor presented in this thesis.

Given these considerations, the concept of the *channel efficiency* proves to be a helpful tool to compare different 3D sensors under information theoretical aspects and to estimate their performance.

Remark 2: The basic concept of evaluating optical 3D sensors under information theoretical aspects is being practiced at the Erlangen Institute of Optics for a long time. Instead of the introduced *channel efficiency*, the authors of past publications [21, 24, 25] propose a *different* measure for the evaluation: The *information efficiency* is the relation of the *realized mutual information of the 3D dataset* $H(Z_{mq})$ and the 2D channel capacity ($\triangleq \mathcal{M}_{2D}$):

$$\eta_{inf} = \frac{H(Z_{mq})}{\mathcal{M}_{2D}} \tag{2.34}$$

What is the difference between $H(Z_{mq})$ *and* $\mathcal{M}_{3D}$ *and why does this thesis propose the channel efficiency for the information theoretical evaluation?*

As mentioned, $\mathcal{M}_{3D}$ defines the *highest possible* amount of information to be transferred by a 3D height map. The amount of information which is transferred *in reality*, is presented by $H(Z_{mq})$. Commonly, this amount is much lower than $\mathcal{M}_{3D}$. The reason is the *redundancy* of the measured surfaces. 'Everyday objects', such as human faces or technical parts, display considerably small surface variations, leading to a high central peak of the power spectrum in the Fourier domain and thus, to a high redundancy. A redundancy reduction can be achieved by proper source encoding. Some approaches are described in [21, 24, 25]. For example, the standoff-distance, omnipresent in each 3D coordinate, can be removed/suppressed by intrinsically measuring the *slope* of an object surface instead of the absolute height. For a *completely uncorrelated signal*, the power spectrum is constant vs. the spatial frequency. For an object with such a (very unlikely) shape, $H(Z_{mq})$ would be very close to $\mathcal{M}_{3D}$.

This reveals that the *information efficiency* is strongly dependent on the measured object surface, whereas the *channel efficiency* is not. The author of [21] proposed a trick to estimate $H(Z_{mq})$ for a single example measurement to calculate the information efficiency: The acquired 3D dataset is quantized with the noise level and compressed by a zip-algorithm. The remaining amount of data serves as an upper estimation for $H(Z_{mq})$. This is a reasonable approach if an *'example dataset'* - *representative* for the desired measurement task - exists.

However, the sensor presented in this thesis should be able to measure a broad variety of surface types. A planar sheet of paper should be measurable as well as a 'hedgehog-like' object with spikes of statistically distributed lengths. In this case, it seems reasonable to judge the sensor by the object types it *should be able to measure* instead of concrete example datasets. This ability is expressed by the *channel efficiency*.

2.3.4 Information Theoretical Optimization

The 'recipe' for the information theoretical optimization of a 3D sensor is mainly given in Eq. (2.32):

- ***Optimized optical setup!***
  A poor optical setup decreases the transferable bandwidth (see also Eq. (2.29)). In addition, if the sampling theorem is not satisfied, the optical setup delivers false data from aliasing. Moreover, a good optical setup is the basic requirement for low noise (see below).
- ***As few raw images as possible!***
  Each (parallel or sequentially) acquired camera image, necessary for the calculation of the 3D image, requires additional channel capacity. From this point of view, *single-shot principles* are favorable, since they do not rely on image sequences.
- ***High density of independent 3D points!***
  At a first glance, this point contradicts the above one: Due to the discussed lack of information (three unknowns in one camera image), single-shot principles commonly display a very poor 3D data density.
  The development of a single-shot principle with a high density of independent 3D points is the central point of this thesis.
- ***Low noise!***
  Noise reduces $\mathcal{M}_{2D}$ as well as $\mathcal{M}_{3D}$. Only with the exception of averaging over a large series of images, precise 3D data can principally not be generated from noisy 2D data. Thus, both *SNR*s in Eq. (2.32) have to be large with a considerably larger *SNR* for the 3D data.

The discussion of the last section also reveals how the mutual information, that actually passes the channel, can be maximized:

- ***Proper source encoding!***
  Redundancy should be minimized by proper source encoding. Prominent examples are the measurement of the slope instead of the distance (e.g. Photometric Stereo,

Phase measuring deflectometry [28, 29]) or the pre-distortion of a projected pattern/signal for the precise measurement of an object with approximately known surface (e.g. Interferometry with reference, 'single-shot deflectometry' [30]).

What is information theoretical optimization?

Due to the definitions of this thesis, an optical 3D sensor is ***information theoretically optimized****, if it displays a* ***large channel efficiency****, meaning that the sensor has the* ***potential to generate a large amount of 3D information*** *from the provided amount of 2D information.*

For the most triangulation based principles a sensor with $\eta \approx 33\%$ is estimated ***as good as information theory allows****.*

References

1. G. Häusler, W. Heckel, Light sectioning with large depth and high resolution. Appl. Opt. **27**(24), 5165–5169 (1988)
2. G. Häusler, "Lecture: Technische Optik und optische 3D-Sensoren." www.optik.uni-erlangen.de/osmin. Script Accessed 1 Mar 2017
3. F. Willomitzer, "FlyFace – 3D-Gesichtsvermessung mittels "Flying Triangulation"," Diploma Thesis, University Erlangen-Nuremberg (2010)
4. ISO, "Guide to the Expression of Uncertainty in Measurement" (2008). www.bipm.org/utils/common/documents/jcgm/JCGM_100_2008_E.pdf. Accessed 15 May 2017
5. C. Faber, "New Methods and Advances in Deflectometry," Dissertation, University Erlangen-Nuremberg (2012)
6. Creaform, "Go Scan 3D product website." www.goscan3d.com/en/features. Accessed 15 May 2017
7. S.W. Hell, J. Wichmann, "Breaking the diffraction resolution limit by stimulated emission: Stimulated-emission-depletion fluorescence microscopy." Opt. Lett. **19**(11), 780–782 (1994)
8. Lord Rayleigh Sec. R. S., On the theory of optical images, with special reference to the microscope. Philos. Mag. Ser. **5**(42), 167–195 (1896)
9. K. Strehl, "Aplanatische und fehlerhafte Abbildung im Fernrohr." Zeitschrift für Instrumentenkunde 362–370 (1895)
10. G. Häusler, S. Ettl, *"Limitations of Optical 3D Sensors," Book chapter in 'Optical Measurement of Surface opography'* (Springer, Heidelberg, 2011), pp. 23–48
11. J.W. Goodman, *Introduction to Fourier optics* (Roberts and Company Publishers, 2005)
12. J.C. Dainty (ed.), *Laser Speckle and Related Phenomena*, vol. 9 (Springer Science and Business Media, 2013)
13. G. Häusler, "Speckle and Coherence", in *Encyclopedia of Modern Optics*, vol. 1 (Academic Press, Elsevier, 2004), pp. 114–123
14. A. Philippi, *"Speckle - Ein Praktikumsversuch für Fortgeschrittene," Thesis* (University Erlangen-Nuremberg, Zulassungsarbeit, 1996)
15. G. Häusler, "Ubiquitous coherence - boon and bale of the optical metrologist," Speckle Metrology, Trondheim. Proc. SPIE **4933**, 48–52 (2003)
16. R.D. Dorsch, G. Häusler, J.M. Herrmann, Laser triangulation: Fundamental uncertainty in distance measurement. Appl. Opt. **33**(7), 1306–1314 (1994)
17. J. Goodman, "Speckle: Friend or Foe?" in *Proceedings of the 3rd International Topical Meeting on Optical Sensing and Artificial Vision, AIP 1537*, pp. 5–7 (2013)

18. Z. Yang, "Ein miniaturisierter bewegungsunempfindlicher 3D-Sensor," Master Thesis, University Erlangen-Nuremberg (2009)
19. R. Lampalzer, "Physikalische Grenzen von optischen 3D-Sensoren mit phasenmessender Triangulation," Dissertation, University Erlangen-Nuremberg (2003)
20. J. Habermann, "Statistisch unabhängige Specklefelder zur Reduktion von Meßfehlern in der Weißlichtinterferometrie," Diploma Thesis, University Erlangen-Nuremberg (2002)
21. C. Wagner, "Informationstheoretische Grenzen optischer 3D-Sensoren," Dissertation, University Erlangen-Nuremberg (2003)
22. C.E. Shannon, A mathematical theory of communication. Bell Syst. Tech. J. **27**(3), 379–423 (1948)
23. C.E. Shannon, Communication in the presence of noise. Proc. IRE **37**(1), 10–21 (1949)
24. C. Wagner, G. Häusler, Information theoretical optimization for optical range sensors. Appl. Opt. **42**(27), 5418–5426 (2003)
25. O. Hybl, "Information-Efficient White-Light Interferometry," Dissertation, University Erlangen-Nuremberg (2013)
26. R. Johannesson, "Informationstheorie - Grundlagen der (Tele-)Kommunikation" (Addison-Wesley, Studentlitteratur, 1992)
27. M. Tur, K.C. Chin, J.W. Goodman, When is speckle noise multiplicative? Appl. Opt. **21**(7), 1157–1159 (1982)
28. G. Häusler, M.C. Knauer, C. Faber, C. Richter, S. Peterhänsel, C. Kranitzky, K. Veit, "Deflectometry: 3D-Metrology from Nanometer to Meter." in *Fringe 2009* (Springer, Heidelberg, 2009)
29. C. Faber, E. Olesch, R. Krobot, G. Häusler, "Deflectometry challenges interferometry - the competition gets tougher!." in *Proceedings of SPIE 8493, Interferometry XVI: Techniques and Analysis, 84930R* (2012)
30. H. Liang, E. Olesch, Z. Yang, G. Häusler, Single-shot phase measuring deflectometry for cornea measurement. Adv. Opt. Technol. **5**(5–6), 433–438 (2017)

Chapter 3
State of the Art: The Basic Principles of Optical 3D Metrology

This chapter introduces a selection of *basic principles* for the acquisition of three-dimensional surface data of macroscopic objects with scattering surfaces. As depicted in Chap. 2, most of the applied principles in this regime rely on triangulation. During the following descriptions, special attention is paid to advantages and drawbacks in reference to the desired specs *speed, precision, high feature resolution and data density* as well as how the *correspondence problem*, introduced in Sect. 2.1.1, is solved. Of course, sensors which apply the basic methods in their 'pure form' are rarely found in the literature or in the market. Most available sensors exploit a modification of the 'pure form'. It is clear, that not all of these modifications can be discussed here. However, some single-shot principles will be discussed in more detail.

A second 'state of the art section' (Sect. 10.1) will deal with a few selected methods, that use a similar idea or seem to be strongly related to the methods and approaches introduced in this thesis. To understand similarities and differences, these methods will be discussed after the explanations of Chap. 5.

3.1 Preamble: Stereo Vision

The human eye is an extremely paradigmatic invention of nature. It serves as the most important link between the human brain and the outside world and is the archetype for all classical imaging systems. Even more important for this thesis: The whole human vision system is more than the sum of its parts. Two eyes, paired with the well-defined geometrical arrangement in the head, enable the ability of 3D vision by triangulation.

Hence, the human vision system is the *archetype* of all existing triangulation methods. The underlying principle is called ***'passive stereo-photogrammetry'***, or in short ***'passive stereo'***. Two imagers (cameras, eyes) observe the object under a

F. Willomitzer, *Single-Shot 3D Sensing Close to Physical Limits and Information Limits*, Springer Theses,
https://doi.org/10.1007/978-3-030-10904-2_3

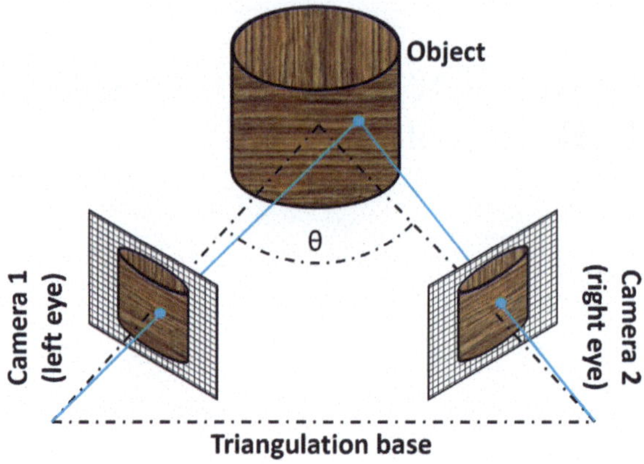

Fig. 3.1 Basic principle of passive stereo photogrammetry: An object is observed by two imagers under a triangulation angle. The object surface is sufficiently structured in order to identify correspondences

certain triangulation angle θ. In order to find the correct correspondences in the two images, the object surface has to be sufficiently structured, e.g. by intrinsic texture, see Fig 3.1. If this cannot be ensured, the surface is artificially structured by pattern projection (***'active stereo'***). The pros and cons of different patterns will be discussed later. In order to span a defined triangulation base, it is important that the system has *at least* ***two*** *calibrated components*! This means (see Fig 3.2):

- one calibrated camera and one calibrated projector or
- two calibrated cameras and one uncalibrated projector just for 'synthetic texture' projection (avoids elaborate projector calibration).

This is very fundamental and plays an important role for the rest of this thesis. Nearly all triangulation methods used for optical 3D metrology are *active* methods and follow these rules.

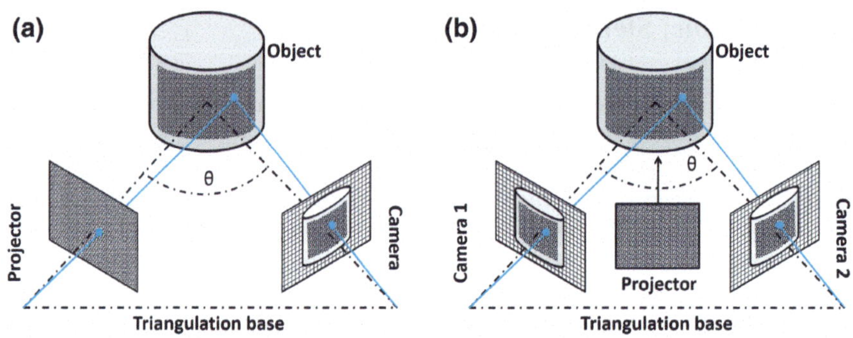

Fig. 3.2 Basic principle of active stereo photogrammetry: The object is artificially structured by projection. In order to span a triangulation base, two options are possible: **a** One calibrated camera and one calibrated projector. **b** Two calibrated cameras and one projector

Subsequently, some active methods will be described. Active stereo photogrammetry will be shortly discussed again in Sect. 3.3.1, when '*Microsoft Kinect One*®' is introduced.

3.2 Sequential Methods

3.2.1 Phase Measuring Triangulation

Phase measuring triangulation (PMT) is the *archetype of all sequential triangulation methods*. It is still used with slight modifications in a majority of 3D sensors due to its prominent features: *high precision* and *high density of 3D data* (see below). The basic principle was first described by Srinivasan, Liu and Halioua in 1984 [1]: A pattern, composed of sinusoidal fringes is projected onto the object surface, observed with a camera under the triangulation angle. Due to the depth variations of the surface, the fringes are deformed in the camera image (see Fig. 3.3). Hence, the observed *phase* $\varphi(x', y')$ of the fringes encodes the depth z (modulo 2π). The observed intensity in each pixel (x', y') can be generally expressed as

$$I(x', y') = A(x', y') + B(x', y') \cdot \cos(\varphi(x', y')) \,. \tag{3.1}$$

Equation (3.1) contains ***three unknowns***: The (wanted) phase $\varphi(x', y')$ as well as $A(x', y')$ and $B(x', y')$, which contain information about the unknown bias illumination and object reflectivity. *Three unknowns* require at least *three equations* to be solved. These equations are generated via a *temporal phase shift* procedure [2]: $F \geq 3$ exposures are taken. Between these exposures the projected phase is shifted by $\Delta\varphi = \frac{2\pi}{F}$. The observed intensity $I_s(x', y')$ in the sth shift is

$$I_s(x', y') = A(x', y') + B(x', y') \cdot \cos(\varphi(x', y') - \varphi_s) \,, \qquad \varphi_s = (s-1)\frac{2\pi}{F} \,. \tag{3.2}$$

Finally $\varphi(x', y')$ can be evaluated [2] by

$$\varphi(x', y') = \arctan \frac{\sum_{s=1}^{F} I_s(x', y') \cdot sin\varphi_s}{\sum_{s=1}^{F} I_s(x', y') \cdot cos\varphi_s} \tag{3.3}$$

For $F = 4$ ('*four-phase shift*'), the algorithm becomes very simple and, in addition, insensitive to second order nonlinearities:

$$F = 4 \quad \Rightarrow \quad \varphi(x', y') = \arctan \frac{I_2(x', y') - I_4(x', y')}{I_1(x', y') - I_3(x', y')} \tag{3.4}$$

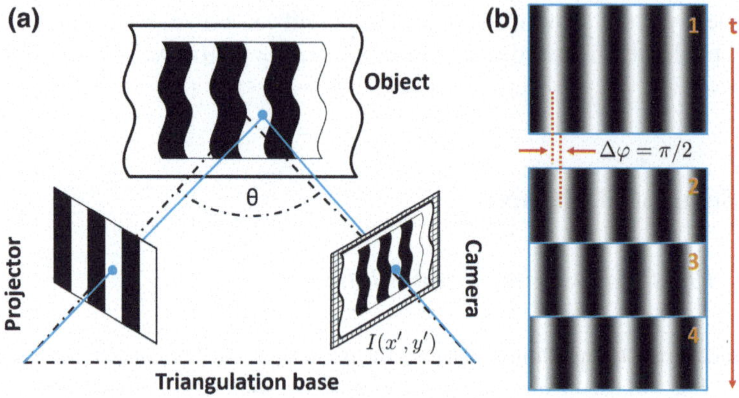

Fig. 3.3 Basic principle of phase measuring triangulation (PMT). **a** A sinusoidal pattern is projected onto the object. The phase of the fringes in the camera picture deciphers the object height and is evaluated by temporal phase shifting. **b** *'Four-phase shift'* as an example for phase shifting: Four subsequent images are acquired. Between each image, the phase of the projected pattern is shifted by $\Delta\varphi = \pi/2$

However, in practice commonly many more sequential images ($F \approx$ 8–30) are required. This is related to the profound *ambiguity problem*. As discussed, $\varphi(x', y')$ can be only evaluated by modulo 2π. For a precise 3D evaluation, the fringe frequency should be high, which produces ambiguities to a large extent. The problem is solved for example by successive phase shifts with different fringe frequencies ('multi-frequency phase shift') [3] or by projecting an additional sequence of a binary 'gray-code[1]' [5], which enables 'phase unwrapping'. A detailed overview about different strategies is given in [6–8].

Although PMT is not a single-shot principle, it represents the 'gold standard' of 3D metrology. This has mainly two reasons: First, it is very precise. A measurement uncertainty down to several μm can be achieved, which, for example, enables measurements of dental casts or coins [2, 9] with sufficient precision. Second, PMT is a *purely local* method. Each camera pixel produces a 3D point without relying on information from neighbored pixels. This is why PMT is often described as 'pixel-dense' with $\rho_{3D} = 100\%$. However, 100% can not fully be reached, since the sampling theorem has to be satisfied. And indeed, even PMT displays artifacts at sharp edges, where the sampling theorem is violated.

Nevertheless, the channel efficiency of 'very good' PMT approaches reaches high values around $\eta_{PMT} \approx$ 20–30% (see Sect. 6.5). Single-shot methods, described in Sect. 3.3, have commonly a much lower channel efficiency. This thesis intends to develop a single-shot principle that reach a channel efficiency similar to the good values of η_{PMT}.

[1]The 'gray-code' is a cyclic permutation binary code named after Bell Labs researcher Frank Gray [4]; also known as 'reflected binary code'.

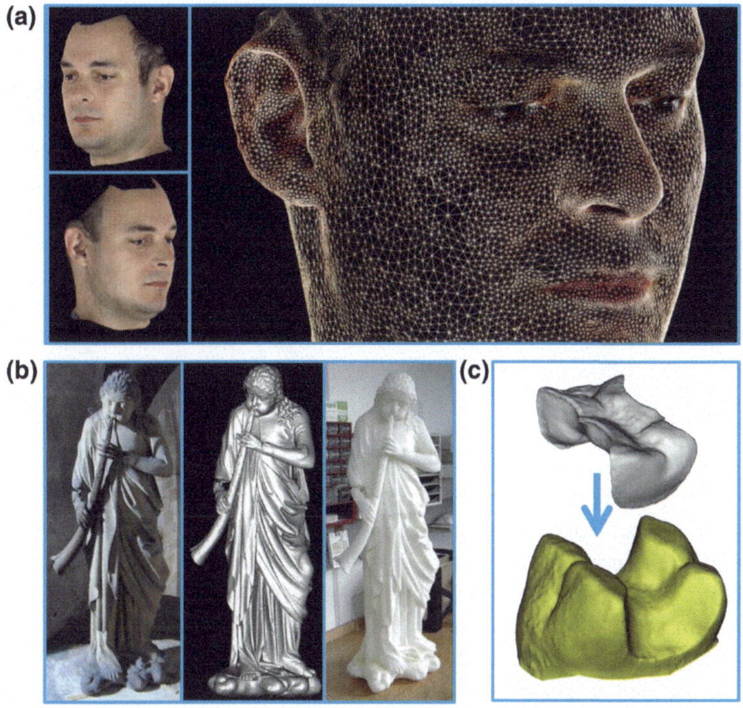

Fig. 3.4 Measurement examples from PMT sensors developed at the Institute of Optics in Erlangen. **a** Facial scan: 3D model from two different positions and triangular mesh. **b** The 'Trombone Angel' of the Bamberg cathedral. Left to right: original, 3D model, printed copy. **c** Treated tooth and dental inlay

The Erlangen Institute of Optics investigated, developed and constructed PMT sensors since the late 90s with focus on medical applications, such as measurements of the human body, face or teeth [2, 9–12]. Figure 3.4 shows a few examples. The sensors are commercialized by the spin-off '3D-Shape GmbH[2]' [13].

3.2.2 *Other Sequential Methods*

In simplified terms one can say that *all sequential methods follow a similar basic approach*: In each coordinate (pixel) on the camera chip, a *temporal sequence* of intensity values deciphers a coordinate on the projector chip (or the chip of the second camera) to obtain the correspondence. Eventually, a depth value is calculated via triangulation of these two coordinates. This procedure does not necessarily require

[2]www.3d-shape.com.

a sinusoidal pattern. Principally any random pattern (within the limits of the sampling theorem) can be projected.

Sequential methods, that do not use the classical PMT approach, commonly do this all for the same purpose: *speed up the total measurement time for one 3D frame* in order to enable also measurements of moving objects. This is necessarily related to fast hardware. Due to the improvements in chip technology, cameras with high frame rates are affordable nowadays. Mostly, the projector is the bottleneck. For an 8 bit pattern-image with 256 gray values, a DLP projector requires 256 mirror switching cycles until the next pattern can be projected. Depending on the measured scene, this can be too long to project a sequence of patterns without object motion in between. This 'projector-problem' is solved in different ways. Some illustrating methods are discussed in the following.

Binary Phase Shifting

By projecting a binary pattern the projector speed can be significantly increased. The projector pixels (mirrors on DLP) only have to switch *once* per picture in *on*- or *off-state*. Hence, a binary image can be projected 256 times faster than an 8 bit gray image (sufficient luminous intensity assumed).

Multiple options for the design of binary patterns are available. Projecting a binary stripe pattern provokes phase errors in the evaluation. This can be solved by defocusing the projector or by algorithmic methods [14]. The latter may involve low pass filtering. Another possibility is to use dithered patterns [15]. Other methods are illustrated, e.g. in [6–8].

Temporal Correlation of 'Random' Patterns

The discussed 'projector problem' can also be solved by using a projector, that does *not* require electronic switching. The group of Kowarschik in Jena has been following this approach for many years. The basic setup is comparable to the setup of Fig. 3.2b: Two cameras and one projector, which principally does not have to be calibrated into the system. A pseudo random, band-limited pattern is projected, which constantly moves over the object surface. This can be done, for example, by moving a slide with the imprinted pattern on an overhead-projector or by projecting the pattern onto a wobbling mirror [16]. With a high-frame rate camera, a series of images $I_{t_1}, \ldots, I_{t_n}$ is acquired in a short time. Each camera pixel $(x', y')_{C_1}$ on camera 1 delivers a temporal sequence of gray values $[I_{t_1}(x', y')_{C_1}, \ldots, I_{t_n}(x', y')_{C_1}]$. The correct correspondence to pixel $(x', y')_{C_2}$ on camera 2 can be found by (temporal) correlation of the gray value sequence $[I_{t_1}(x', y')_{C_2}, \ldots, I_{t_n}(x', y')_{C_2}]$ with the sequence from camera 1.

A robust result requires $n \approx 15$ to 30 temporal images [17]. Since all images should be acquired in a short time (compared to the object motion), one of the main challenges in this approach is the need of *bright light sources*. A tricky solution is the use of moving laser speckles from a rotating diffuser as projected pattern [17].

As PMT does, the described method delivers nearly independent 3D points in each camera pixel. If a reduced feature resolution and 3D data density is sufficient (e.g. at very smooth objects), the number of required images can be reduced by observing the temporal evolution of pixel-clusters (e.g. 5×5 windows). This 'spatiotemporal correlation' [18] is a representative example for the free exchange between space and time when dealing e.g. with correlation information.[3]

The authors of [21, 22] use a similar criterion as described above to obtain correspondence in the camera images, but project an aperiodic sinusoidal fringe pattern. The fast varying pattern is generated with a rotating slit mask in front of a high power light source (*GOBO-Projector*).

Spatial Codification

Spatial codification approaches are commonly used in single-shot principles for correlation. In sequential methods as PMT, similar approaches can be used to replace time consuming multi-frequency phaseshifts. In [23], a sinusoidal pattern is superimposed by a two-dimensional codification pattern. Although the principle is fast, the spatial codification consumes space bandwidth. Other multi-shot approaches exploiting spatial codification are given e.g. in [6].

3.3 Single-Shot Methods

In this thesis, the term 'single-shot' is defined in that way, that ***no temporal sequence*** of exposures is required for the acquisition of one 3D frame. This implies an essential advantage: If the exposure time is short enough, single-shot principles are ***motion-robust***, which enables a free movement of the sensor and the object during the measurement.

Single-shot principles, that do not exploit additional modalities (color, polarization, frequency limitations,...), can ***never*** deliver 'pixel-dense' data ($\rho_{3D} = 100\%$) in one 3D frame! This is due to the discussed *three unknowns*, which cannot be calculated from one camera picture. A common way to overcome this dilemma is to project a *binary signal*. With proper thresholding, variations in surface reflectivity or background illumination have nearly no effect on the evaluation of the signal.

However, binary signals cannot be projected 'pixel-dense', which is why all single-shot data display a certain sparsity, often concealed by interpolation in the final 3D dataset.

But not all single-shot principles rely on binary projection. Principles, using patterns with different gray values, are able to reach higher data densities, albeit they are more sensitive to texture or reflectance variations.

In the following, several single-shot principles will be described. As this is relevant for the remaining part of the thesis, the discussions will be more detailed and also consider sensors, which are commercially available.

[3] In other words: The *Space-time-bandwidth-product* $STBP$, also known as 'information cube' is 'incompressible', see e.g. [19, 20].

3.3.1 Microsoft Kinect One® (Active Stereo)

The '*Microsoft Kinect One®*' 3D sensor (released in 2010) was originally developed for an interactive controlling of computer games. The user (player) is positioned in front of the sensor and controls his avatar by body movements (e.g. playing virtual tennis by moving his arm). Since also fast movements have to be detected, a single-shot approach is applied.

The basic setup of the sensor is comparable to Fig. 3.2a: One camera and one calibrated projector. An additional color camera allows the acquisition of color texture. The measurement principle is based on active stereo photogrammetry. A pseudo random, speckle-like dot-pattern is projected onto the object, see Fig. 3.5a, b. The wavelength used is in infra-red, which allows for a simultaneous color texture acquisition.

Since the single dots are indistinguishable, the correspondences between projected pattern and camera image have to be found by comparing the *spatial arrangement of dots* in a defined correlation window (e.g. the red window in Fig. 3.5a, b). This consumes space-bandwidth, which is lost for high feature resolution of the 3D image, as displayed in Fig. 3.5c.

For smooth objects and the originally intended gaming application, this might be tolerable. For metrology applications, like the measurement of mechanical parts, this fact is unfavorable, as all high frequencies, including edges, are smoothed out due to low pass filtering.

Nevertheless, 'Kinect One®' became one of the 'standard sensors' in 3D vision. Since the public release of the Microsoft Kinect® SDK (Software Development Kit), the sensor found its way into scientific applications; mainly in medicine, art preservation or some fields of computer vision. Last but not least this is because of the large distribution, the low price ($\sim$ 100 USD) and the (obviously uncritical) user community.

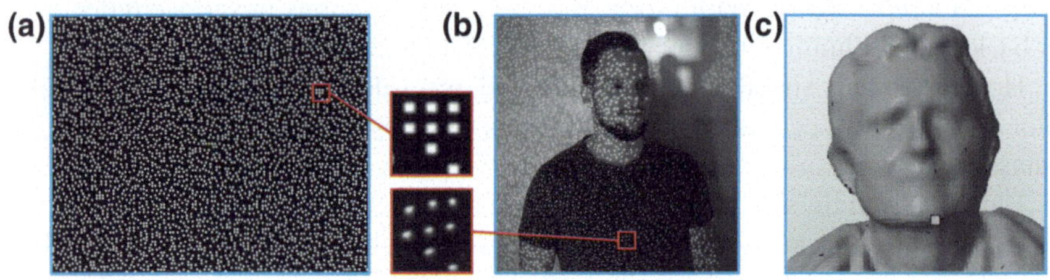

Fig. 3.5 Basic principle of 'Microsoft Kinect One®'. A pseudo random, speckle-like pattern is projected onto the object. The projected wavelength is not visible for the human eye (IR). **a** Part of the projected pattern [27]. **b** Pattern projected on a person. Red frames: Correspondences between (**a**) and (**b**). **c** Measurement result of a human face

As ‘Microsoft Kinect One®’ is a commercially sold product, the basic principle is only described in very few scientific publications. For more detailed information, it is recommended to review the related patent applications [24–26] as well as some technical websites [27–30].

In 2014, Microsoft released the second generation of its ‘Kinect®’ sensor, which is based on a *‘Time-of-Flight’* approach (see Sect. 3.5.1).

3.3.2 Fourier Transform Profilometry

Fourier transform profilometry (FTP) is one of the few principles that is able to acquire a ‘pixel-dense’ point cloud in one single-shot. As discussed, this is not possible until something else has to be condoned. In the case of FTP, this is a bandwidth limitation of the object surface. The basic principle was first described by Takeda et al. in 1983 [31]. A narrow sinusoidal fringe pattern is projected onto the object surface, as shown in Fig. 3.6a. The acquired camera image is Fourier-transformed (Fig. 3.6b). If the object surface is ‘smooth’ enough, the spectrum Q_1 around the carrier frequency f_0 can be directly separated from the other spectra in the Fourier domain. After filtering and Fourier-back-transformation, the phase variation $\Delta\varphi(x, y)$ can be separated from the unwanted amplitude variation and the object height $z(x, y)$ is calculated.

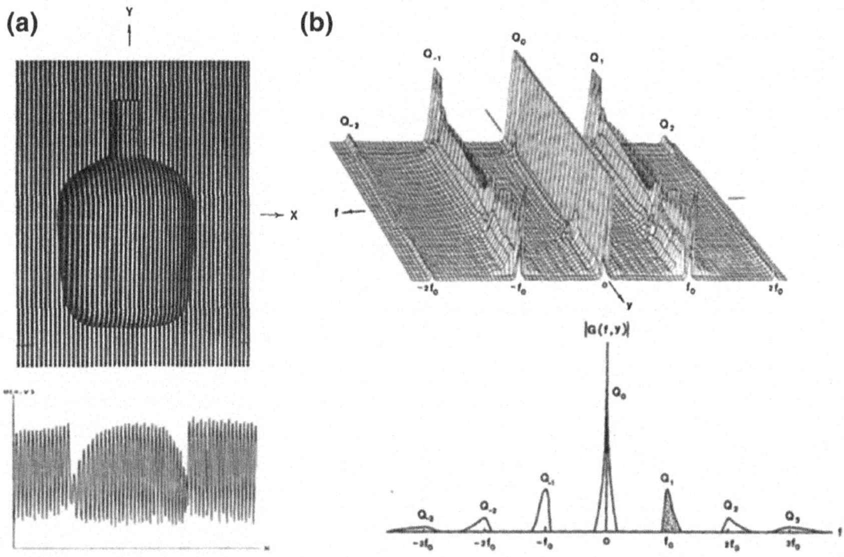

Fig. 3.6 Original figures, taken from [31]: **a** Object with deformed projected pattern. Cross section below. **b** Related Fourier spectrum. Spectrum Q_1 around the carrier frequency f_0 has to be separable from other spectra

Where is the limit? $\mathcal{Q}_1$ has to be separable in the Fourier domain. According to [31], this is the case if

$$\left|\frac{\partial\varphi(x,y)}{\partial x}\right|_{max} < \frac{1}{3}\cdot 2\pi f_0 \,. \tag{3.5}$$

This restricts the maximal allowed object slope to

$$\left|\frac{\partial z(x,y)}{\partial x}\right|_{max} < \frac{1}{3}\cdot\frac{l_0}{d}\,, \tag{3.6}$$

where l_0 is the stand-off distance and d is the triangulation base. The triangulation angle θ can be defined as $\tan\theta = d/l_0$. For a common triangulation angle of $\theta = 10°$, the object slope has to be smaller than 1.9. This is only satisfied by relatively smooth objects. In the meantime several methods are developed to overcome this limitation. Some of them e.g. rely on the subsequent acquisition of a second image, which makes the principle not single-shot anymore. Another interesting modification that enables the measurement of surface discontinuities is the *'frequency-multiplex FTP'* [32]. An overview of different methods is given e.g. in [33].

3.3.3 *Flying Triangulation (Multi-Line Triangulation with Registration)*

In line triangulation or 'light sectioning' a narrow straight light line is projected onto the object surface. Since the line direction is perpendicular to the triangulation-plane (see Fig. 3.7), the line is deformed in the camera picture due to the surface height variations. From this deformation, the object shape along the line is calculated.

Similar to PMT for multi-shot principles, line triangulation is very paradigmatic for single-shot principles because it creates a nearly perfect 3D profile: As discussed, the object surface is scanned in line direction with best possible feature resolution, since no neighborhood information is exploited. Perpendicular to the line, only a small amount of spatial bandwidth is consumed by sub-pixel interpolation of the line intensity maximum (see Sect. 6.1). Therefore, light sectioning is able to measure high object frequencies, e.g. of mechanical parts. And indeed, a multitude of single-line triangulation (SLT) scanners (often called 'laser scanners' because of the light source used) is available for industrial applications – mainly to be mounted on robot-arms or above band-conveyors.

The serious drawback of line triangulation is its *small data density*. 3D data can be only generated along the projected line. For a 1 Megapixel camera ($1000 \times 1000\, pix$) this relates to $\rho_{SLT} = 0.1\%$.

Projecting multiple lines, like shown in Fig. 3.7, helps to increase the data density. However, common multi-line triangulation (MLT) sensors are *only* able to project $\sim$10 lines relating to $\rho_{MLT} \approx 1\%$. The reason for this small amount of projected

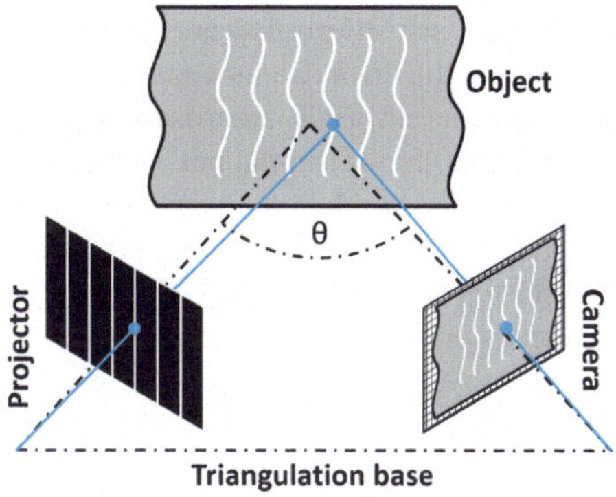

Fig. 3.7 Basic principle of (multi-) line triangulation. One or more narrow light lines are projected onto the object surface, which is observed under the triangulation angle. Due to the surface height variations, the line is deformed in the camera image. The object shape along the line is calculated from this deformation

lines is the *correspondence problem*: Straight, narrow lines are indistinguishable! Codification approaches, which can help to distinguish the lines, are undesired in order to preserve high feature resolution.

A workaround for this problem was presented by the Erlangen Institute of Optics in 2009: ***'Flying Triangulation'*** [34–38]. The method is based on multi-line triangulation with ~10 narrow lines *plus* a sophisticated registration algorithm. The basic principle is demonstrated in Fig. 3.8: In order to fill the gaps in the final 3D image, a temporal series of sparse 3D profiles is acquired, while the sensor is moved by hand

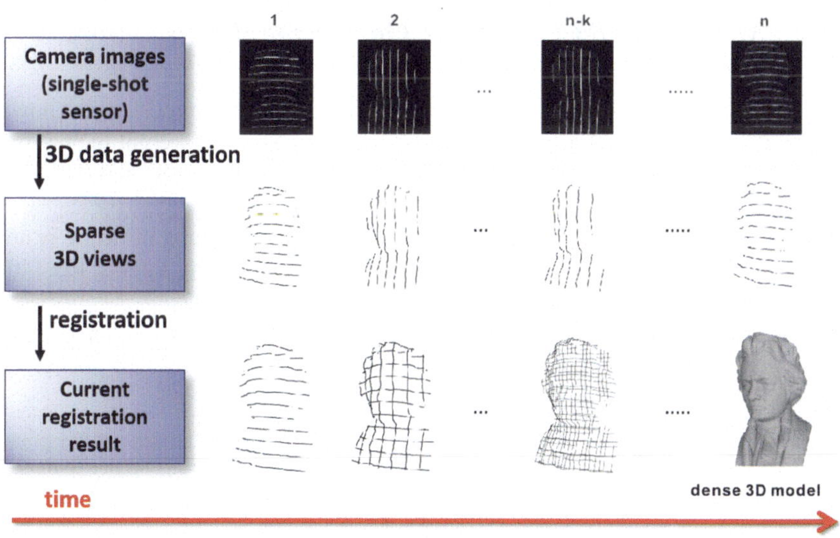

Fig. 3.8 Basic principle of 'Flying Triangulation'

around the object surface. The sparse 3D profiles are subsequently registered to each other in order to compensate for the unknown movement. The registration algorithm is based on an 'iterative closest point' (ICP) approach. It relies on the existence of common measured surface points in two subsequent 3D profiles, so-called 'temporal corresponding points'.[4] Flying Triangulation ensures the occurrence of 'temporal corresponding points' by an alternating projection of horizontal and vertical line patterns. The registration works *'on line'* and is displayed as interactive user feedback on the computer screen during the measurement. Color texture information is acquired with an additional color camera by flash exposure. 3D models measured with Flying Triangulation are shown in Sect. 4.3. Videos of full measurement sequences can be seen on [39].

Although listed at single-shot principles, Flying Triangulation is not a single-shot principle in the common sense: It relies on a *temporal series of exposures* for a full 3D scan. However, each 3D profile for itself is acquired in one single-shot, which makes the principle robust against motion. During the measurement a free (rigid) movement of object *and* sensor is allowed.

Remark: As described, Flying Triangulation presents a *workaround* for the profound correspondence problem in multi-line triangulation. This thesis presents methods and approaches to *solve* the correspondence problem. However, at the beginning of this thesis, the starting point of all considerations was the improvement of Flying Triangulation in terms of reliability and data density per single-shot.

In the next chapter, a detailed description of the correspondence problem in multi-line triangulation is given, using the example of Flying Triangulation. Moreover, several sensors and innovations for Flying Triangulation, developed within the framework of this thesis, are discussed.

3.3.4 Artec (Spatial Coded Multi-Line Triangulation)

In the last years, the *'Artec'*[5] company emerged to one of the most well known producers for hand held 3D scanners. 'Artec' scanners are omnipresent in newspapers or TV documentations about archeology, forensic applications or 3D printing applications.

As the scanners are commercially sold products, the underlying principle is not published in detail. However, information can be found in relating patents, like [40] and -for the experienced audience- in product videos, like [41, 42].

From this references it becomes obvious, that 'Artec' uses a special kind of spatial coded multi-line triangulation: The correspondence problem is solved by

---

[4]In this case the word 'corresponding' refers to *temporal correspondence*: The same surface point, measured with a *moving* camera in subsequent exposures. In this thesis, 'correspondence' is mostly used for *spatial correspondence*: The same surface point, measured simultaneously from different directions.

[5]www.artec3d.com.

thickness-variations of the projected lines. From line to line the thickness is changed in different periods (see Fig. 3.9), which makes the lines distinguishable. Exploiting the epipolar-geometry of the sensor additionally helps to resolve ambiguities.

Figure 3.9 deals with the projected pattern. Figure 3.9a is taken from the patent [40], and shows an *'illustration of a representative light structure'* (pattern). Figure 3.9b consist of two screenshots from a product video [41] on the official 'Artec' YouTube-channel. The pattern, projected by the scanner, is clearly visible and resembles the pattern of Fig. 3.9a.

To distinguish between the lines, the binary thickness-code has to be read by the computer. This requires neighborhood information, since a line segment of a certain length has to be visible in the camera picture. The consumed space bandwidth is lost for the resolution of high object frequencies. Features, smaller than the required segment, cannot be measured in line direction.

The codification also consumes space bandwidth in the perpendicular direction due to the thickness variations. This limits the number of projected lines and therefore the data density of the sensor principle. According to Fig. 3.9a, the projected pattern is assumed to consist of ~80 coded lines. With this an upper bound for the data density can be calculated to $\rho_{Artec} < 8\%$ for a 1-Megapixel camera ($1000 \times 1000\ pix$). The real data density is estimated smaller, since only *independent* points are considered. The value above has to be divided by the minimal line segment length (unknown) that ensures the identification of the 'code'. For a reasonable value of 10–15 pix, the data density shrinks to $\rho_{Artec} < 0.8\%$. As already known from other sensors, the low data density is concealed to a large extent by interpolation and high resolution color texture, which is acquired during an additional flash exposure with a color camera.

Although not indicated by the references, it is not impossible that 'Artec' also uses a short temporal sequence of exposures or 3D frames to increase the data density or to use the information for correspondence. The drawbacks of this approach have already been discussed.

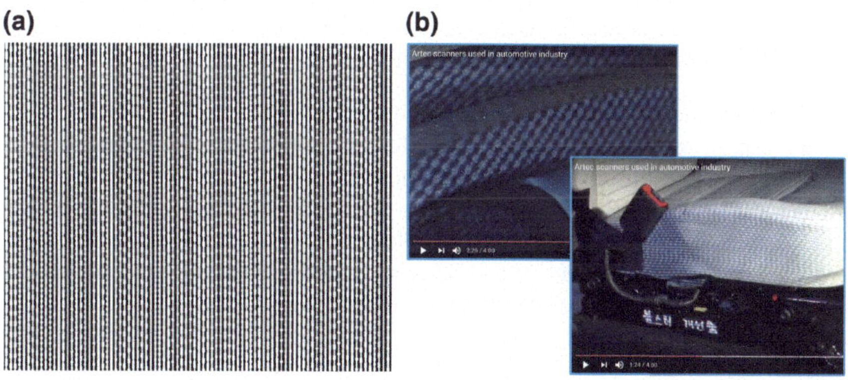

Fig. 3.9 Pattern projected by Artec sensors. **a** Original figure taken from patent [40], showing a representative pattern to be projected. **b** Screenshots from a product video [41]. The projected pattern is clearly visible and resembles the pattern shown in (**a**)

The approach of coded multi-line triangulation is not exclusively followed by Artec. Similar approaches with similar advantages and drawbacks are e.g. [43, 44]. The Erlangen Institute of Optics investigated a line codification approach as well [45–48] to increase the number of projected lines in Flying Triangulation. Here, the thickness of the projected lines remains constant, but the intensity profile is coded with a sinusoidal pattern of different frequencies to distinguish the lines. Such intensity codings can help to save 'space' on the camera chip, but reduce the SNR in the camera image.

3.3.5 Color Coded Triangulation

Instead of their shape, lines can also be distinguished by their *color*. The color coding methods follow exactly this approach. Color cameras are required for the measurement. Most cameras acquire color texture information with a 'Bayer pattern' - a pattern of small color filters, mounted in front of the camera chip. Each small filter is positioned directly in front of one pixel. One 2×2 'elementary cell' of the filter pattern consists of two green, one red and one blue filter (rarely: two yellow, one magenta and one cyan).[6] This has two major drawbacks: First, the effective chip resolution of the camera is reduced. Each image point contains information from at least three pixels. Most cameras are able to export a red, green and blue channel picture in the original chip resolution. However, these pictures are interpolated. Second, pixels behind the green, red and blue filters have different SNRs. Depending on the light source, the transmitted intensity strongly varies between the different filters. If e.g., a line intensity maximum has to be calculated, this is a source of noise. However, expensive 'three-chip-cameras' can solve the discussed problems to a certain extend.

Another problem of the color coded triangulation is the acquisition of the intrinsic object color texture. Since object texture is not 'spectrally pure', the reflected color is strongly dependent on the used light source. This can be seen as an artifact in the 3D model. In many sensor principles an additional flash exposure is used for the object texture acquisition. The lost single-shot ability is the price.

Several approaches for color coded triangulation can be found in literature.
The most intuitive approach is to project different color coded lines. The authors of e.g. [49, 53–55] use this method. But this is not all. In many cases, the pattern *period* of a line color is varied, too, like shown in Fig. 3.10a. This helps to reduce the number of projected colors, because additional correlation information is created. However, distinguishing lines by pattern periods requires neighborhood information. Spatial bandwidth is consumed. The 3D models display larger artifacts at edges, were no neighborhood information is available.

[6] The reason, that green appears twice as often as red and blue is related to the spectral emission curve of the sun and the spectral sensitivity of the human eye (photopic V_λ-Curve). Both have their maximum in green. For a transformation of color images to gray images, it is in many cases sufficient just to look at the green channel.

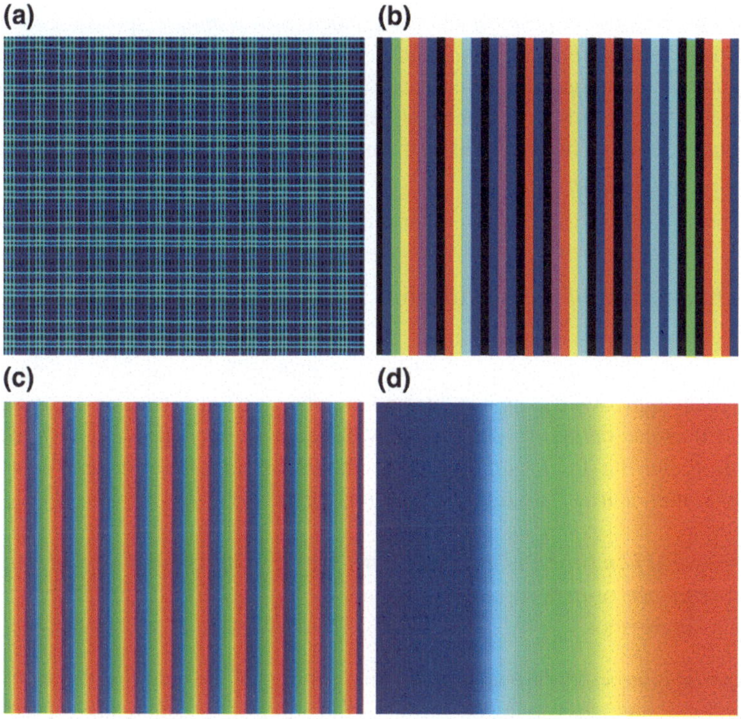

Fig. 3.10 Examples for projected patterns in color coded triangulation approaches. **a** Color coded lines with varied pattern period, taken from [49] (© 2009 IEEE). **b** Color coded stripes, taken from [50] (© 2002 IEEE). **c** Color coded and phase shifted sinusoidal fringes, taken from [51]. **d** Continuous color spectrum, as it is used in [52]

Other approaches use the projection of color coded stripes, proposed e.g. by [50, 56, 57]. In order to correctly distinguish between the projected colors, the projected stripes have to have a certain width (see Fig. 3.10b). To calculate 3D data, the edges between two stripe colors are triangulated. Edge detection always involves differentiation operations, which are very noise sensitive. Note as well, that some approaches like [50] again rely on information of the extended neighborhood.

Another option to be discussed is inspired by PMT-phase shifting (see Sect. 3.2.1): Instead of projecting three phase shifted sinusoidal patterns subsequently, it is also possible to use a different color for each phase shifted pattern (red, green, blue) and project everything in one picture, as shown in Fig. 3.10c. Methods that follow this idea are described e.g. in [51, 58]. Note that the projection of several phase shifts in one colored image does not include the solution of the ambiguity problem, which still requires measurements with multiple fringe-frequencies (multi-shot, as in PMT).

The last option to be discussed is 'Rainbow Triangulation' [52, 59], invented by Häusler et al. in 1993. The projected 'pattern' consists of a continuous spectrum of wavelengths ('rainbow'), that can be generated by the dispersion of a white light beam through a prism. Such a projection, like shown in Fig. 3.10d can be seen as a pattern of *densely* projected lines, each coded with a *unique* spectral color. This means, that the principle delivers *'pixel-dense'* 3D data in each single-shot *without correspondence ambiguities*. However, projecting a bright spectrum is technically elaborate and requires high power broadband light sources. Nevertheless, the principle might become more important in the near future, by applying newly developed supercontinuum lasers.

3.4 Discussion of Triangulation Principles

In the last two sections, it became obvious that there is no 'perfect' triangulation sensor. Each approach has its benefits and drawbacks. It also became clear, that the decision, if a sensor property is *really* a drawback is strongly dependent on the desired application. For the discussion of the preceding examples, the *beneficial properties of triangulation sensors* are divided in *three groups*:

- ***No temporal bandwidth restrictions***. This excludes all methods relying on a temporal sequence of exposures for the generation of one 3D frame. Such methods are not single-shot.
- ***No spatial bandwidth restrictions*** (within the limits of the sampling theorem). This excludes all methods relying on spatial codification, color codification or requiring relatively 'smooth' object surfaces. Such methods are not able to measure high frequency components of the object surface.
- ***No ambiguity problem***. Such methods are principally able to reach a high density of 3D data. However, if extensive spatial codification is exploited for the solution of the ambiguity problem, the 3D data density might be reduced.

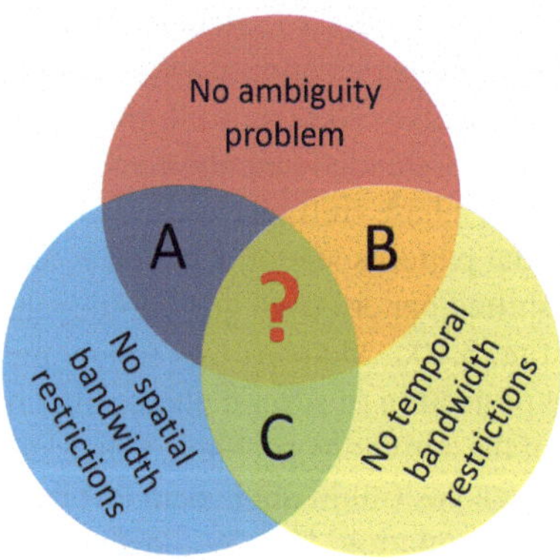

A	Phase Measuring Triangulation	Section 3.2.1
	Temp. Correlation of 'Random' Patterns	Section 3.2.2
B	(Active) Stereo Photogrammetry	Section 3.3.1
	Fourier Transform Profilometry	Section 3.3.2
	Spatial Coded Multi-Line Triangulation	Section 3.3.4
	Color Coded Triangulation	Section 3.3.5
C	(Multi-) Line Triangulation	Section 3.3.3

Fig. 3.11 Categorization of triangulation principles

Having in mind this classification and looking at the discussed sensor principles again, it is remarkable, that each sensor principle displays *not more than* ***two*** of the described properties at the same time. Figure 3.11 shows a graphical representation. The main question is:

'Why is there no sensor, that displays all three properties at once?'

'Is this fundamental?'.

The answer is: *'yes'*.

This answer was the motivation and starting point of the presented thesis.

In Chap. 5 an approach is introduced, which indeed displays all three properties at once. As mentioned, this would not be possible without purchasing something.

3.5 Other Methods, Not Based on Classical Triangulation

As discussed, most optical measurement principles for the acquisition of macroscopic objects with scattering surfaces are based on triangulation. However, a few principles, enabling this task, are not. For the sake of completeness, the most important approaches are discussed in this section.

3.5.1 Time-of-Flight

Besides triangulation a distance can be also measured by the *'Time-of-Flight' (ToF)*. ToF-based approaches for distance measurement are already well known from Radar, Sonography (Ultrasound) or Interferometry. ToF 3D cameras, having similar measurement tasks like the system presented in this thesis, are mostly based on the so-called PMD chip technology (*Photonic Mixing Device*), invented by Schwarte [60]: A *temporally intensity-modulated* light beam flood illuminates the object and is reflected from its surface. In each pixel of the PMD chip, the phase shift $\Delta\varphi$ between outgoing and incoming wave is measured, which deciphers the object distance z by

$$z = \frac{1}{2} \cdot \frac{c}{f} \cdot \frac{\Delta\varphi}{2\pi}, \tag{3.7}$$

where f is the modulation frequency. Commonly, a sinusoidal modulation is used, but also other modulations, like separated pulses, are possible. The basic principle of a PMD chip can be understood as follows: Each pixel consist of at least two light-sensitive modulation gates connected with capacitive elements (containers). As a function of the (known) wave modulation, the gates are switched and the generated charges are guided to the corresponding containers. By comparing the amount of charges in the containers at the end of an integration interval, the phase of the incoming wave can be calculated. With a 'dark frame' (exposure without active illumination), ambient light is compensated and the object texture can be acquired.

In the meantime, ToF cameras gained a growing importance for 3D applications, like in industry, medicine or multimedia. For example, the second generation of Microsoft's Kinect® sensor is based on ToF, as already mentioned in Sect. 3.3.1. The success of the technique is mainly based on four advantages. First, the PMD chips became affordable nowadays and the price will even decrease in the future. Second, the technique is very fast. Frame rates up to 160 fps can be achieved. Third, the camera delivers a 3D point in *each* pixel and the pixel resolution of the chips is steadily increasing. Last, the principle is a coaxial technique, which eliminates occlusion artifacts.

In current systems a depth resolution of $\sim$10 mm is reached. This is already sufficient for many applications, but ***far away*** from the precision of triangulation approaches.

Although it is often said that ToF-cameras *'deliver a depth value in each camera frame'*, ToF is not a single-shot principle in the classical sense as defined at the beginning of this thesis. Note that the ambiguity problem is also present in ToF-approaches. The 'ambiguity interval' for ToF-cameras is $\Delta Z = \frac{1}{2}\frac{c}{f}$ and represents the 'unique measurement range'. For common modulation frequencies of 25–100 MHz, ΔZ varies from 6–1.5 m. Similar to PMT an extension can be reached by multi-frequency approaches (variation of modulation frequency).

3.5.2 Holography

Holography was invented by Gabor in 1948 [61] and awarded with the Nobel Prize in 1971. The basic idea is to record a three-dimensional light field on a two-dimensional photosensitive material (or later: camera chip). The full information of the light field can be reconstructed later when the original source is no longer present.

The light field is generated e.g. by illumination with a plane wave, which is scattered from the surface of an object under test. For recording, the deformed wavefront is superimposed with a 'reference beam' (mostly generated with a beam splitter) with the same properties like the original wave for illumination. The photo plate records an interference pattern consisting of relative phase differences between 'object beam' and 'reference beam'. For evaluation, the photo plate is illuminated with a 'reconstruction beam', that has the same properties like the 'reference beam'. This enables the reconstruction of the full surface information of the object.

The common application of holographic methods in metrology is microscopy. Only a few publications deal with applications comparable to this thesis. In [62–64] a holographic method is used for the 3D metrology of human faces and other macroscopic objects. The approach still uses analog holograms. The 3D reconstruction is performed by so-called 'Hologram-Tomography' [65]: A ground glass is moved along the reconstructed hologram step by step and the 'slices' with maximum contrast are recorded. This technique is elaborate and leaves room for improvement in terms of precision. The noise in the data is mainly caused by speckle noise.

The reconstruction techniques of digital holography [66] can help to simplify the procedure. However, this works only for relatively small viewing angles, which is why digital methods are mostly used for the measurement of small objects.

Another possibility to exploit holographic methods for 3D metrology is the combination with stereo photogrammetry [67–69]. In this approach, holograms, recorded from two stereo cameras, are used (after speckle removal) for the depth calculation via triangulation. The calibration of the setup is simplified considerably.

3.5.3 Photometric Stereo

Although it uses angle dependent information, *Photometric Stereo* (PS) is not a triangulation approach in the classical sense. Instead of absolute heights, PS measures the gradient of a surface by observing the object under changing illumination conditions. The method is based on one fundamental assumption: The amount of light, which is reflected from a surface to an observer, is dependent on the surface orientation relative to observer and light source.

The original technique, pioneered by Woodham [70], assumes surfaces with *Lambertian scattering*, *uniform albedo*, and *point-like light sources* with *known positions*. The object is placed under the ($\geq$3) light sources, which are preferably arranged around the camera [71, 72]. Camera pictures are taken while the light sources are switched on and off alternately, creating different shading images from the object surface. By solving an inverse equation system, the surface normal can be calculated in each camera pixel. A single-shot approach, based on a similar technique, is ***'Shape from Shading'***, developed by Horn [73].

Although Photometric Stereo or Shape from Shading is commonly not used for high precision shape measurements, the principle is simple, fast and information efficient [72, 74]. As the mentioned methods are very sensitive to local slope variations, the detection of surface imperfections is one of the key applications. For the measurement of absolute height data, a known 'starting point' onto the object surface is required, since only gradient data are measured. The main weakness of the original principle is the need of 'Lambertian' surfaces, which makes it inapplicable for many 'everyday situations'. By applying some techniques, developed by the Computer Vision community, the method can be generalized to more realistic situations, like textured, 'non-Lambertian' or even specular surfaces, see e.g. [75–77] and many others. As usual, such approaches require additional sources of information.

Another -very intuitive- way to use Photometric Stereo for the metrology of **all** kinds of surfaces is the approach followed by the group of Adelson [78, 79]: Instead of *directly* measuring the surface, the object is *imprinted* in a fine powder or on a powdered surface of a special gel-pad. The resulting detailed surface-imprint displays all properties required for standard high quality Photometric Stereo approaches. However, the method is not contactless anymore and may damage or influence the surface under test.

3.6 Conclusion

The goal of this thesis is the development of a *single-shot* 3D sensor, that acquires *precise* 3D models with a *high density* of independently measured 3D points, also displaying *small surface features*. Given the overview of the previous chapter, it becomes obvious that *'single-shot'* and *'high data density and feature resolution'* are contradictory words for the current state of the art.

Only one single-shot principle displays high feature resolution and data density, even though in *only one direction*: *(multi-) line triangulation*. Although the data density, perpendicular to the line direction, is very poor, the data density and feature resolution *along* the lines is the *best possible* to be reached within the limits of the sampling theorem. The absence of extensive spatial coding prevents low-pass filtered data.

Because of these prominent features, the sensor principle, developed in this thesis is based on a ***multi-line triangulation*** approach. To reach the desired *high data density* in one single-shot, the profound *'indexing problem'* has to be solved.

References

1. V. Srinivasan, H. Liu, M. Halioua, Automated phase measuring profilometry of 3-D diffuse objects. Appl. Opt. **23**, 3105–3108 (1984)
2. M. Gruber, G. Häusler, Simple, robust and accurate phase-measuring triangulation. Optik **89**(3), 118–122 (1992)
3. H.O. Saldner, J.M. Huntley, Temporal phase unwrapping: application to surface profiling of discontinuous objects. Appl. Opt. **36**(13), 2770–2775 (1997)
4. F. Gray, "Pulse Code Communication," US Patent No. US 2,632,058 (1953)
5. I. Ishii, K. Yamamoto, K. Doi, T. Tsuji, High-speed 3D image acquisition using coded structured light projection, in *IEEE/RSJ International Conference on Intelligent Robots and Systems* (2007), pp. 925–930
6. J. Salvi, J. Pages, J. Batlle, Pattern codification strategies in structured light systems. Pattern Recognit. **37**(4), 827–849 (2004)
7. J. Salvi, S. Fernandez, T. Pribanic, X. Llado, A state of the art in structured light patterns for surface profilometry. Pattern Recognit. **43**(8), 2666–2680 (2010)
8. J. Geng, Structured-light 3D surface imaging: a tutorial. Adv. Opt. Photon. **3**, 128–160 (2011)
9. G. Häusler, M. Benz, P. Ettl, S. Karbacher, X. Laboureux, R. Lampalzer, K. Veit, E. Nkenke, Optical 3D-sensing and CAD/CAM for medicine and art conservation, in *Seventh International Symposium on Laser Metrology Applied to Science, Industry, and Everyday Life*, International society for optics and photonics (2002)
10. X. Laboureux, G. Häusler, Localization and registration of three-dimensional objects in space - where are the limits? Appl. Opt. **40**(29), 5206–5216 (2001)
11. G. Häusler, P. Andretzky, H.-P. Habermeier, S. Karbacher, I. Laszlo, X. Laboureux, N. Schön, K. Veit, Optische 3D-Sensoren helfen dem Restaurator. Das moderne Museum - Die Vorträge der MUTEC **99**, 6–14 (2001)
12. S. Karbacher, S. Seeger, G. Häusler, Refining triangle meshes by non-linear subdivision, in *IEEE Proceedings of Third International Conference on 3-D Digital Imaging and Modeling* (2001), pp. 270–277
13. "3D-Shape GmbH, FaceSCAN^{3D}, BodySCAN^{3D} product sheet," www.3d-shape.com/produkte/face_e.php. Accessed 15 May 2017

14. S. Zhang, High-resolution 3D profilometry with binary phase-shifting methods. Appl. Opt. **50**(12), 1753–1757 (2011)
15. W. Lohry, S. Zhang, High-speed absolute three-dimensional shape measurement using three binary dithered patterns. Opt. Exp. **22**, 26752–26762 (2014)
16. M. Grosse, M. Schaffer, B. Harendt, R. Kowarschik, Fast data acquisition for three-dimensional shape measurement using fixed-pattern projection and temporal coding. Opt. Eng. **50**(10), 100503 (2011)
17. M. Schaffer, M. Grosse, R. Kowarschik, High-speed pattern projection for three-dimensional shape measurement using laser speckles. Appl. Opt. **49**(18), 3622–3629 (2010)
18. B. Harendt, M. Grosse, M. Schaffer, R. Kowarschik, 3D shape measurement of static and moving objects with adaptive spatiotemporal correlation. Appl. Opt. **53**(31), 7507–7515 (2014)
19. A.W. Lohmann, R.G. Dorsch, D. Mendlovic, C. Ferreira, Z. Zalevsky, "The space-bandwidth product, applied to spatial filtering and holography," Research Paper RJ-438 (IBM San Jose Research Laboratory) (1967), pp. 1–23
20. A.W. Lohmann, Space-bandwidth product of optical signals and systems. J. Opt. Soc. Am. A **13**, 470–473 (1996)
21. S. Heist, P. Kühmstedt, A. Tünnermann, G. Notni, Theoretical considerations on aperiodic sinusoidal fringes in comparison to phase-shifted sinusoidal fringes for high-speed three-dimensional shape measurement. Appl. Opt. **54**(35), 10541–10551 (2015)
22. S. Heist, P. Lutzke, I. Schmidt, P. Dietrich, P. Kühmstedt, A. Tünnermann, G. Notni, High-speed three-dimensional shape measurement using GOBO projection. Opt. Lasers Eng. **87**, 90–96 (2016)
23. W. Lohry, V. Chen, S. Zhang, Absolute three-dimensional shape measurement using coded fringe patterns without phase unwrapping or projector calibration. Opt. Exp. **22**, 1287–1301 (2014)
24. B. Freedman, A. Shpunt, M. Machline, Y. Arieli, "Depth Mapping using Projected Patterns," US Patent Application No. US 2010/0118123A1 (2010)
25. A. Shpunt, "Depth Mapping using Multi Beam Illumination," US Patent Application No. US 2010/0020078A1 (2010)
26. J. Garcia, Z. Zalevsky, "Range Mapping using Speckle Decorrelation," US Patent No. US 7,433,024 B2 (2008)
27. A. Reichinger, "Kinect Pattern Uncovered," azt.tm's Blog (2011), www.azttm.wordpress.com/2011/04/03/kinect-pattern-uncovered/. Accessed 15 May 2017
28. "Kinect Hacking 103: Looking at Kinect IR Patterns," www.futurepicture.org/?p=116. Accessed 15 May 2017
29. K. Konolige, P. Mihelich, "Technical description of Kinect calibration," wiki.ros.org/kinect_calibration/technical. Accessed 15 May 2017
30. J. Welsh, R. Stewart, "Kinect for Xbox 360: Technology + Two Applications," www.cs.virginia.edu/~jfw3x/cs4501/. Accessed 15 May 2017
31. M. Takeda, K. Mutoh, Fourier transform profilometry for the automatic measurement of 3-D object shapes. Appl. Opt. **22**(24), 3977–3982 (1983)
32. M. Takeda, Q. Gu, M. Kinoshita, H. Takai, Y. Takahashi, Frequency-multiplex Fourier-transform profilometry. Appl. Opt. **36**(22), 5347–5354 (1997)
33. X. Su, W. Chen, Fourier transform profilometry: a review. Opt. Lasers Eng. **35**(5), 263–284 (2001)
34. S. Ettl, O. Arold, P. Vogt, O. Hybl, Z. Yang, W. Xie, G. Häusler, Flying triangulation: a motion-robust optical 3D sensor principle, in *Fringe* (Springer, Heidelberg, 2009), pp. 1–4
35. S. Ettl, O. Arold, Z. Yang, G. Häusler, Flying Triangulation - an optical 3D sensor for the motion-robust acquisition of complex objects. Appl. Opt. **51**(2), 281–289 (2012)
36. F. Willomitzer, S. Ettl, O. Arold, G. Häusler, Flying triangulation - a motion-robust optical 3D sensor for the real-time shape acquisition of complex objects, in *Proceedings of the 3rd International Topical Meeting on Optical Sensing and Artificial Vision, AIP 1537* (2013), p. 19

37. O. Arold, S. Ettl, F. Willomitzer, G. Häusler, Hand-guided 3D surface acquisition by combining simple light sectioning with real-time algorithms (2014). arXiv:1401.1946
38. S. Ettl, Introductory review on 'Flying Triangulation': a motion-robust optical 3D measurement principle. in *Contemporary Physics* (2014)
39. Osmin3D, "YouTube-channel of the OSMIN research group at the University Erlangen-Nuremberg," www.youtube.com/user/Osmin3D. Accessed 15 May 2017
40. N. Lapa, Y. Brailov, "System and Method for Three-dimensional Measurement of the Shape of Material Objects," US Patent No. US 7,768,656 B2 (2010)
41. "Artec scanners used in automotive industry," youtu.be/dyqlY6kHfnE. Accessed 15 May 2017
42. "Car scanning with Artec 3D scanner," youtu.be/H4OaWj62AWk. Accessed 15 May 2017
43. M. Maruyama, S. Abe, Range sensing by projecting multiple slits with random cuts. IEEE Trans. Pattern Anal. Mach. Intell. **15**(6), 647–651 (1993)
44. J. LeMoigne, A.M. Waxman, "Multi-resolution grid patterns for building range maps," Vision '85, in *Applied Machine Vision Conference (ASME)* (1985)
45. Z. Yang, W. Xie, O. Arold, O. Hybl, P. Vogt, K. Bobey, G. Häusler, "A 3D sensor for intraoral metrology," in *Proceedings of the 110th DGaO Conference* (2009), p. 19
46. O. Arold, "Flying Triangulation" - handgeführte optische 3D Messung in Echtzeit, Dissertation, University Erlangen-Nuremberg (2013)
47. F. Willomitzer, "FlyFace – 3D-Gesichtsvermessung mittels "Flying Triangulation"," Diploma Thesis, University Erlangen-Nuremberg (2010)
48. D. Rauen, "Intensitätsmodulierte Linienmuster für FlyTri," Internship Report, University Erlangen-Nuremberg (2011)
49. R. Sagawa, Y. Ota, Y. Yagi, R. Furukawa, N. Asada, H. Kawasaki, "Dense 3D reconstruction method using a single pattern for fast moving object." in *2009 IEEE 12th International Conference on Computer Vision* (IEEE, 2009)
50. L. Zhang, B. Curless, S.M. Seitz, "Rapid shape acquisition using color structured light and multi-pass dynamic programming." in *First International Symposium on 3D Data Processing, Visualization, and Transmission* (IEEE, 2002)
51. P. Huang, Q. Hu, F. Jin, F. Chiang, Color-encoded digital fringe projection technique for high-speed three-dimensional surface contouring. Opt. Eng. **38**(6), 1065–1071 (1999)
52. G. Häusler, D. Ritter, Parallel three-dimensional sensing by color-coded triangulation. Appl. Opt. **32**(35), 7164–7169 (1993)
53. A.O. Ulusoy, F. Calakli, G. Taubin, "Robust one-shot 3D scanning using loopy belief propagation." in *2010 IEEE Computer Society Conference on Computer Vision and Pattern Recognition-Workshops* (IEEE, 2010)
54. J. Pages, J. Salvi, C. Matabosch, "Robust Segmentation and Decoding of a Grid Pattern for Structured Light." in *Iberian Conference on Pattern Recognition and Image Analysis* (Springer, Heidelberg, 2003)
55. R. Sagawa, R. Furukawa, H. Kawasaki, Dense 3D reconstruction from high frame-rate video using a static grid pattern. IEEE Trans. Pattern Anal. Mach. Intell. **36**(9), 1733–1747 (2014)
56. F. Forster, "A high-resolution and high accuracy real-time 3D sensor based on structured light." in *Third International Symposium on 3D Data Processing, Visualization, and Transmission* (IEEE, 2006)
57. C. Schmalz, "Robust single-shot structured light 3D scanning." Dissertation, University Erlangen-Nuremberg (2011)
58. S. Zhang, P. Huang, "High-resolution, real-time 3D shape acquisition." in *IEEE Conference on Computer Vision and Pattern Recognition Workshop, CVPRW'04* (2004)
59. J. Geng, Rainbow three-dimensional camera: new concept of high-speed three-dimensional vision systems. Opt. Eng. **35**(2), 376–383 (1996)
60. R. Schwarte, "Method and Apparatus for Photomixing," US Patent No. US 6,825,455 B1 (2004)
61. D. Gabor, A new microscopic principle. Nature **161**, 777–778 (1948)
62. J.R. Bongartz, "Hochauflösende dreidimensionale Gesichtsprofilvermessung mit kurzgepulster Holographie." Dissertation, University Düsseldorf (2002)

63. D.M. Giel, "Hologram tomography for surface topometry." Dissertation, University Düsseldorf (2003)
64. J.R. Bongartz, D.M. Giel, P. Hering, "Living human face measurements using pulsed holography." in *Holography 2000. International Society for Optics and Photonics* (2000), pp. 303–308
65. A. Thelen, J. Bongartz, D. Giel, S. Frey, P. Hering, Iterative focus detection in hologram tomography. J. Opt. Soc. Am. A **22**(6), 1176–1180 (2005)
66. U. Schnars, C. Falldorf, J. Watson, W. Jüptner, *Digital Holography and Wavefront Sensing* (Springer Book, Berlin, 2015)
67. M. Grosse, J. Buehl, H. Babovsky, A. Kiessling, R. Kowarschik, 3D shape measurement of macroscopic objects in digital off-axis holography using structured illumination. Opt. Lett. **35**(8), 1233–1235 (2010)
68. H. Babovsky, M. Grosse, J. Buehl, A. Kiessling, R. Kowarschik, Stereophotogrammetric 3D shape measurement by holographic methods using structured speckle illumination combined with interferometry. Opt. Lett. **36**(23), 4512–4514 (2011)
69. T. Pitkäaho, T.J. Naughton, "Stereo vision based approach for extracting features from digital holograms", in *Fringe 2009* (Springer, Heidelberg, 2009)
70. R. Woodham, Photometric method for determining surface orientation from multiple images. Opt. Eng. **19**(1), 191139 (1980)
71. M.C. Knauer, C. Richter, G. Häusler, 3D sensor zoo - Species and natural habitats. Laser Technik J. (1), 33–37 (2006)
72. F. Wolfsgruber, C. Rühl, J. Kaminski, L. Kraus, G. Häusler, R. Lampalzer, E.-B. Häußler, P. Kaudewitz, F. Klämpfl, A. Görtler, "A simple and efficient optical 3D-Sensor based on "Photometric Stereo" (UV-Laser Therapy)", in *Fringe 2005* (Springer, Heidelberg, 2006), pp. 702–706
73. B.K.P. Horn, *"Obtaining Shape from Shading Information"* (MIT Press, 1989), pp. 121–171. Book chapter in "Shape from Shading"
74. C. Wagner, G. Häusler, Information theoretical optimization for optical range sensors. Appl. Opt. **42**(27), 5418–5426 (2003)
75. K. Ikeuchi, Determining surface orientations of specular surfaces by using the photometric stereo method. IEEE Trans. Pattern Anal. Mach. Intell. **3**(6), 661–669 (1981)
76. H.D. Tagare, R.J. de Figueiredo, A theory of photometric stereo for a class of diffuse Non-Lambertian surfaces. IEEE Trans. Pattern Anal. Mach. Intell. **13**(2), 133–152 (1991)
77. D.B. Goldman, B. Curless, A. Hertzmann, S. Seitz, Shape and spatially-varying brdfs from photometric stereo. IEEE Trans. Pattern Anal. Mach. Intell. **32**(6), 1060–1071 (2010)
78. M.K. Johnson, E.H. Adelson, "Retrographic Sensing for the Measurement of Surface Texture and Shape." in *IEEE Conference on Computer Vision and Pattern Recognition, CVPR 2009* (2009)
79. M.K. Johnson, F. Cole, A. Raj, E.H. Adelson, Microgeometry capture using an elastomeric sensor. ACM Trans. Graph. (TOG) **30**(4) (2011)

Chapter 4
Introducing the Problem

In this chapter the so-called *'indexing problem'* of multi-line triangulation will be introduced, motivated by the example of the sensor principle Flying Triangulation (see Sect. 3.3.3 and [1–5]). The indexing problem is the main reason for the low efficiency of current multi-line triangulation approaches.

As discussed, Flying Triangulation was originally designed as a *workaround* on the indexing problem. However, this yields only if a special class of objects is measured. As soon as an object is not suited to the sensor specifications, Flying Triangulation is also heavily affected by false line indexing. Similar problems occur if the sensor is operated by an inexperienced user. In the second part of this chapter, several Flying Triangulation sensors developed before and during this thesis are introduced. Related methods, that try to avoid indexing errors, are discussed.

4.1 How to Distinguish Undistinguishable Lines?

The drawbacks of spatial codification were discussed in the last chapter. It became clear that only the projection of narrow, uncoded lines preserves the resolution of high object frequencies for line triangulation. But how to distinguish these lines from each other for a correct evaluation?

The most intuitive approach is: *counting* from left to right. However, this only works for a very limited class of object surfaces [4, 6, 7]. In 'daily life' applications line segments or even whole lines are not visible in the camera image. This can be caused by occlusion effects, low surface reflectivity or objects considerately smaller than the measurement field.

Due to this fact the problem of distinguishing lines is solved in Flying Triangulation by so-called *'regions of uniqueness'*. This approach requires the definition of a *fixed* measurement depth ΔZ. Figure 4.1 explains the principle. For simplicity, a telecentric setup with $N_L = 5$ projected lines is illustrated. The lines are projected into the camera's field of view under the triangulation angle θ. If a horizontal plane

F. Willomitzer, *Single-Shot 3D Sensing Close to Physical Limits and Information Limits*, Springer Theses,
https://doi.org/10.1007/978-3-030-10904-2_4

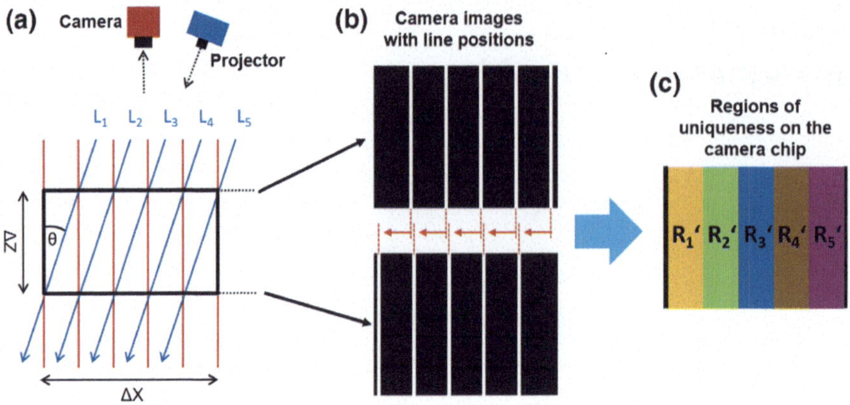

Fig. 4.1 Correct line indexing with 'regions of uniqueness'. **a** Lines projected under the triangulation angle θ in the camera's field of view (width ΔX, given by the used optics) and resulting unique measurement depth ΔZ. **b** 'Shift regions' on the camera chip following from the vertical movement of an imaginary plate within ΔZ. **c** Resulting 'regions of uniqueness' on the camera chip

would be moved vertically (along z-direction) inside the field of view, the images of the projected lines would be shifted horizontally in the camera images, as illustrated by Fig. 4.1b. The *'unique measurement range'* ΔZ represents the vertical distance where the plane can be moved without an overlap of the 'shift regions'. These 'unique shift regions' are the *'regions of uniqueness'*, shown in Fig. 4.1c. In other words: Inside ΔZ each pixel on the camera chip corresponds only to **one** line number (from now on: *'line index'* n).

Surface points measured inside ΔZ by intersection with the N_L projected lines $L_1, \ldots, L_{N_L}$ are automatically imaged onto the corresponding areas $R'_1, \ldots, R'_{N_L}$ on the camera chip, where the line index n is determined [8]. This is shown in Fig. 4.2. With known n the calibration-transformation K, that maps each point (x, y, z) in space to pixel coordinates (x', y') onto the camera chip

$$\begin{pmatrix} x' \\ y' \end{pmatrix} = K \begin{pmatrix} x \\ y \\ z \end{pmatrix} \tag{4.1}$$

can uniquely be inverted to K^{-1}. With the pixel coordinates of the measured signal (x'_s, y'_s) and the determined index n, the 3D data (x_e, y_e, z_e) of the object surface can be evaluated via

$$\begin{pmatrix} x_e \\ y_e \\ z_e \end{pmatrix} = K^{-1} \begin{pmatrix} x'_s \\ y'_s \\ n \end{pmatrix}. \tag{4.2}$$

However, the described approach for the determination of the index n has drawbacks: The *resolution of 3D data* in x-direction $1/\delta x_{dat} = N_L/\Delta X$, the *triangulation angle*

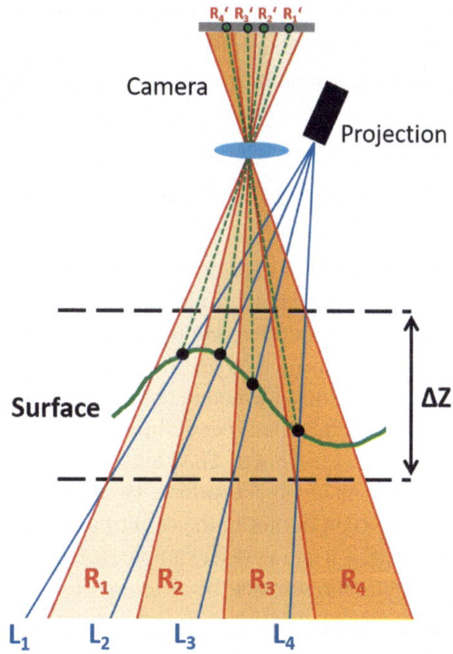

Fig. 4.2 Line indexing inside the unique measurement depth. Each signal produced by intersection of the object surface and the projected lines $L_1, \ldots, L_4$ is imaged into its correct region of uniqueness $R'_1, \ldots, R'_4$ on the camera chip, where the index n is assigned

θ and the *unique measurement depth* ΔZ are strongly related to each other. Presuming *unique* line indexing and looking again at the telecentric example of Fig. 4.1a the data resolution $1/\delta x_{dat}$ can be described as

$$\frac{1}{\delta x_{dat}} = \frac{N_L}{\Delta X} \le \frac{1}{\Delta Z \cdot \tan\theta}\,, \tag{4.3}$$

where the field width ΔX is a system constant determined by the camera chip and optics.

This means: If the number of lines N_L is *increased*, the unique measurement depth ΔZ or the triangulation angle θ (or both) have to become *smaller* in order to guarantee unique line indexing. Equation (4.4) defines an *upper* limit for the product of the *number of projected lines*, the *unique measurement depth* and the *precision*

$$N_L \cdot \Delta Z \cdot \tan\theta \;\le\; \Delta X\,, \tag{4.4}$$

where the relation between precision and triangulation angle is given by Eq. (2.20).

Ideally, a sensor should display a high resolution of 3D data, high unique measurement depth and high precision *at the same time*, which contradicts Eq. (4.4). The design of multi-line triangulation sensors therefore requires certain tradeoffs.

In the special case of Flying Triangulation the situation is even more crucial: In addition to *precise data* the sensor needs a comfortable *large measurement depth*, since it is hand-guided. On the other hand, a *high number of projected lines* is required for robust registration [9].

The next section describes what happens if Eq. (4.4) is violated.

4.2 Indexing Ambiguities in Multi-Line Triangulation

Since the number of projected lines as well as the triangulation angle are fixed in each specific sensor setup, a violation of Eq. (4.4) happens the easiest way if an object point is measured ***outside*** the unique measurement depth ΔZ. This is illustrated in Fig. 4.3: The projected line L_4 intersects the object surface outside of ΔZ. The signal from this point is located in region R_3 and is imaged onto area R'_3 on the camera chip. Since R'_3 corresponds per definition only to line L_3, a wrong line index ($n = 3$ instead of $n = 4$) is assigned. The signal is evaluated like coming from the intersection of line L_3 with an object surface. Hence, a wrong 3D data point is evaluated, which can be seen as an outlier in the resulting 3D model [8].

Figure 4.4 shows two examples for this process measured with two different Flying Triangulation sensors for different object sizes (see next section). In both cases the object is placed *inside* ΔZ, but in front of a wall, which is located *outside* ΔZ. Due to Fig. 4.3 and the descriptions above, the signals, coming from the wall are imaged

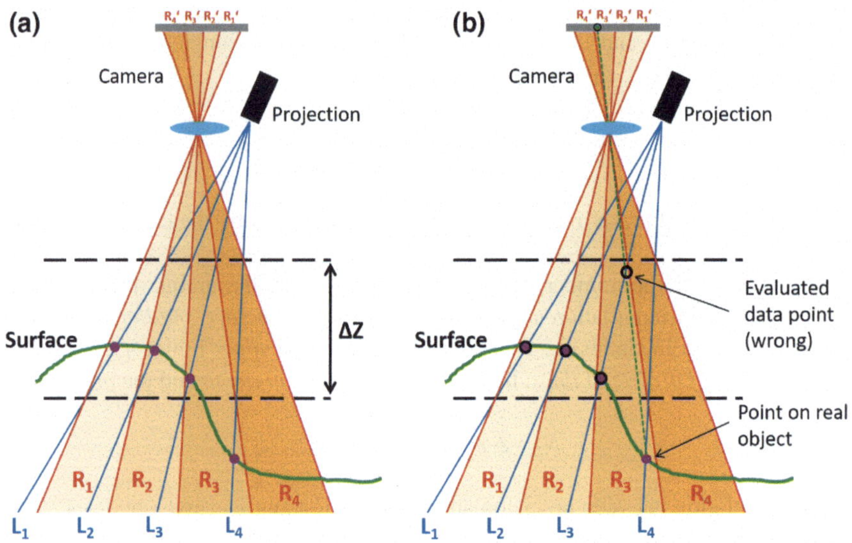

Fig. 4.3 Formation of outliers due to wrong line indexing. **a** A surface point located outside the unique measurement range ΔZ is measured by intersection with Line L_4. **b** The corresponding signal is imaged into the region of uniqueness R'_3 of line L_3. A false index $n = 3$ is assigned and a false 3D data point is evaluated from the signal, which can be seen as outlier in the 3D model

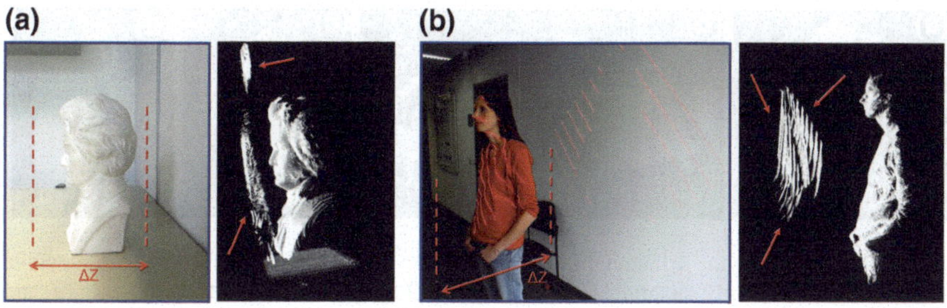

Fig. 4.4 Demonstration of outlier formation in Flying Triangulation for sensors with two different measurement volumes. In both cases, the object is placed in front of a wall, which is outside the unique measurement range ΔZ. The outliers evaluated from the signals of the wall are marked by red arrows in the final 3D data set. **a** Face scanner [10, 11]. **b** Body scanner [12, 13]

in the *false 'regions of uniqueness'* and thus assigned to a false line index n. The generated false 3D data can be seen as outliers (marked by red arrows) in the final 3D dataset.

Note that in both shown examples of Fig. 4.4 the outliers are clearly spatially separated from the correct measured 3D data. If this is the case, outliers could theoretically be removed via *post-processing methods*. In the current Flying Triangulation data acquisition software, an a posteriori outlier removal is only provided 'by hand' using a 3D snipping function. An automatic post-processed outlier removal is not provided. This has a certain reason: The outliers have to be *distinguished* from the correct 3D model. What is clearly visible for a human observer is not trivial to the computer. If the sensor should be able to measure a broad range of object shapes, the class of 'allowed' shapes should not be restricted. This means, that outliers and correct data are not distinguishable in the evaluated 3D model.

However, all 'realistic' objects have surfaces that partially consist only of low frequency components. Here neighbored surface points are clearly connected to each other in 3D space. If only a few separated outliers are produced during the measurement they can be figured out by using 3D neighborhood information. Unfortunately, the amount of outliers in hand-guided Flying Triangulation sensors is strongly user- and object-dependent. For complicated objects, measured by inexperienced users, it may happen that the point cloud of outliers consists of densely connected data and is *even larger* than the point cloud of correct data. In that case, automatic outlier filtering is no longer possible without additional context information about the object surface.

The problem can get even worse: For complicated or large (in relation to the measurement volume) objects it may happen that the outliers even *intersect* the correct 3D model, see Fig. 4.5c, d. In this case, an a posteriori outlier removal from the fully registered 3D model is nearly impossible or requires an immense amount of additional information about the object surface.

A more promising approach is the detection of outliers in each *single* 3D frame. This is also advantageous, as the removed outliers can not disturb the registration algorithm. However, a single 3D frame contains only sparse lines (or line segments) in space, which makes a differentiation between outliers and correct data more difficult.

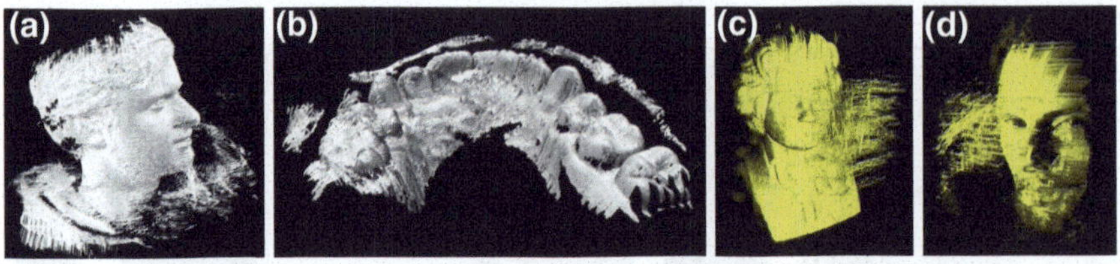

Fig. 4.5 Outliers in Flying Triangulation. An a posteriori outlier removal by hand is only possible for outliers that are clearly separated from the correct 3D data, like it is widely the case for (**a**) and (**b**). For intersecting outliers in (**c**) and (**d**) an outlier removal from the fully registered 3D model is very difficult

The only solution is to ***exploit new sources of information***, which do **not** impose assumptions about the surface shape. A few methods are already described in [9, 11]. For example, one method uses the *width-information* of the imaged lines. The underlying assumption is, that, due to the limited depth of field in projection and observation, a narrow line signal comes more likely from inside ΔZ than a wide line signal. However, the measured line width is also strongly dependent on the volume scattering properties of the object surface, which disturbs the introduced method.

Another method uses an *apparent tilt* of the evaluated 3D data set measured outside ΔZ. This requires a few subsequent images as well as a large amount of neighborhood information and does not work for single 3D points.

An approach based on the calculation of *confidence values* is described in [9] and also implemented in the current version of the Flying Triangulation software. However, this method also uses spatial context information to a large extent.

Hardware-based approaches are also considered. The exploitation of an additional light signal as 'virtual distance control' is discussed in [11]. A mechanical spacer is implemented in the new version of the Flying Triangulation dental scanner (see next section).

The *most promising* approach for correct line indexing in *each* single 3D frame is based on the ***implementation of a second camera*** into the sensor system, which provides additional perspective information. This approach was invented and developed within the framework of this thesis and will be described in detail in the next chapter.

Before the findings of the current chapter are summarized, the sensors developed within the project 'Flying Triangulation' will be shortly presented.

4.3 Flying Triangulation Sensor Zoo

In this section the current sensor-portfolio of Flying Triangulation is introduced. Since Flying Triangulation is freely scalable, many sensors have been developed in the last years, all based on the same simple sensor design shown in Fig. 4.6. Three of the five presented sensors were developed during the time of this thesis under the

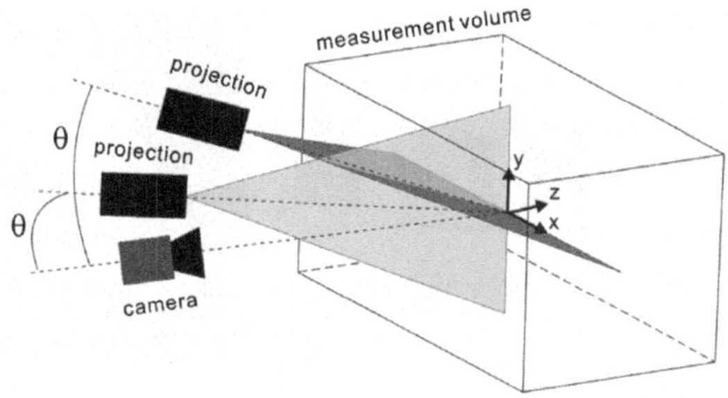

Fig. 4.6 Basic design of a Flying Triangulation sensor: Two projection units alternatingly projecting two orthogonal sparse line patterns onto the object surface, which is observed by a camera under the triangulation angle. Light sources, triangulation angles, stand-off distance and measurement volume vary from sensor to sensor

supervision of the author. One more sensor was developed in the diploma thesis of the author.

The basic measurement principle of Flying Triangulation was introduced in Sect. 3.3.3 and will not be recapitulated here. The sensors are described in the order of their year of development.

The purpose of this section is to demonstrate how the limits introduced by Eq. (4.4) affect the design of multi-line triangulation sensors in practice. The resulting drawbacks and problems are discussed as well.

Dental Scanner (2009)

The high demand for a motion-robust intra oral dental scanner with affordable hardware components was the initial spark of the research project 'Flying Triangulation' in 2008. At that time, sensors for intra oral dental scanning were exclusively based on multi-shot approaches like PMT or focus sensing [14]. Due to the long acquisition times an intra oral measurement without motion artifacts was difficult to achieve even for experienced dentists. Moreover, the related multi-shot techniques require elaborate and expensive hardware like FLCOS displays or precise high speed translation stages. The Flying Triangulation dental scanner works with simple slide projectors, which use a chrome-on-glass mask for pattern generation and a white light LED for illumination (see Fig. 4.7).

Due to the stringent requirements for the measurement of dental casts[1] the sensor is designed to work at the physical limits of triangulation sensors: Several quantities like noise, depth of field and illumination intensity (exposure time) had to be optimized. This optimization puts contradictory constraints on the apertures and triangulation

[1]One of the main application for intra oral dental scanners is the measurement of tooth snags for a precise fitting of the crowns. Bacteria can penetrate the gap between snag and crown for inter-distances larger than 100 μm. Due to tolerances in crown manufacturing the measurement uncertainty of the sensor has to be smaller than 50 μm.

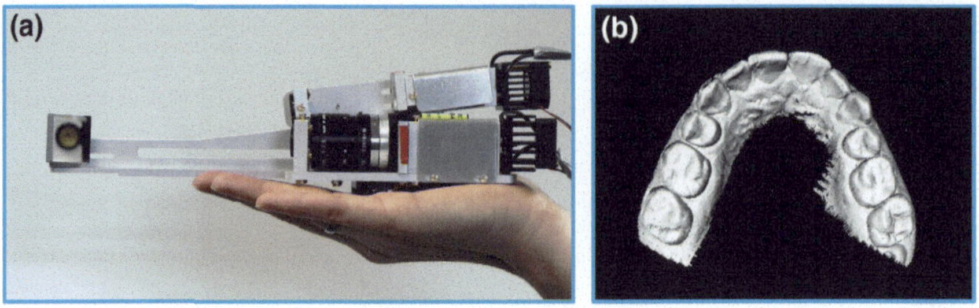

Fig. 4.7 Flying Triangulation dental scanner. **a** Handheld scanner device comprising the components mentioned in Fig. 4.6. **b** Measured 3D model of a full dental cast

angles. The result is a well-defined tradeoff in the optical design, which allows for a 3D data acquisition with a measurement uncertainty better than $\delta z = 30\mu\mathrm{m}$ at 10 vertical resp. 7 horizontal projected lines in a measurement volume of $20 \times 15 \times 15\ \mathrm{mm}^3$ $(\Delta X \times \Delta Y \times \Delta Z)$ with 30 Hz frame rate.

However, there are drawbacks, too: Before the measurement the teeth have to be coated with a spray (TiO_2) in order to get rid of volume scattering. But there is a drawback which is even worse: The small measurement depth provokes indexing ambiguities. During an intra oral hand-guided measurement it is nearly impossible to keep the sensor within the well defined distance range of 15 mm measurement depth. In the special case of dental measurements the resulting outliers are very unfavorable, since a proper a posteriori outlier removal is hard to perform due to the high surface frequency content of the measured object.

Modifications of the sensor setup towards a larger measurement depth would violate the narrow limits of Eq. (4.4). For example, a smaller triangulation angle will increase the measurement uncertainty. A decrease of the number of projected lines is also not an option since the registration algorithm relies on 'corresponding' (crossing-) points. For the discussed sensor setup the number of these points (70 in the very best case) is already at the lower limit for a robust registration [9].

As a workaround to this problem, a hardware-based spacer was applied to the sensor setup. This helps to keep the sensor in the predefined measurement range. Figure 4.8a shows an extra oral measurement of a plaster dental cast with the free

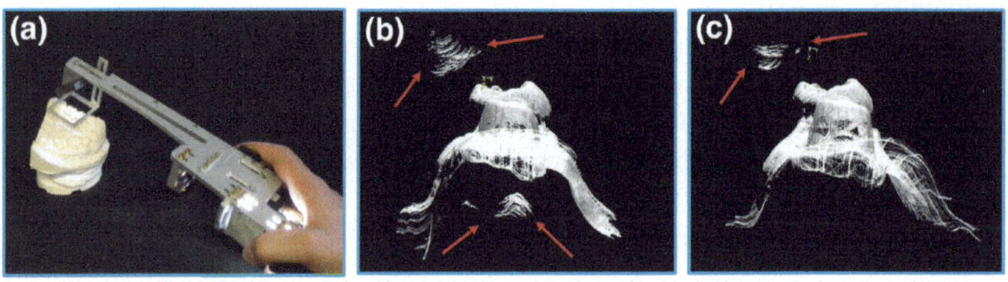

Fig. 4.8 Hardware based approach to reduce indexing ambiguities at dental measurements. **a** Extra-oral measurement of a plaster dental cast with the free hand-guided scanner plus attached spacer. **b** 3D model measured without attached spacer. **c** 3D model measured with attached spacer. Outliers coming from surface points behind the measurement volume can not be prevented by the spacer

hand-guided scanner and spacer. Figure 4.8b, c show corresponding 3D models measured with and without the attached spacer. It becomes obvious that the spacer is not able to prevent all kinds of outliers. Outliers caused by the measurement of surface parts *behind* the measurement volume (upper outliers) are still unavoidable if the object depth is larger than the measurement depth.

For a detailed description of the Flying Triangulation dental scanner see [2, 14, 15].

Face Scanner (2010)

The Flying Triangulation face scanner is originally designed for the 3D measurement of human faces as it is required by many medical applications (esthetic surgery, detection of facial nerve diseases, co-registration of CT-images, ...) or applications in virtual reality. Other medium sized objects like small sculptures, feet or technical parts can be measured as long as their surfaces are rough and offer sufficient structure for the registration algorithm.[2] The face scanner acquires data with 30 Hz along 12 resp. 9 projected lines within a measurement volume of $150 \times 200 \times 100\ \text{mm}^3$ (stand-off distance: 500 mm) and reaches a precision better than 150μm. For the reduction of speckle noise, the scanner setup also comprises slide projectors with white light LEDs and well-designed apertures. Figure 4.9 shows the face scanner 'in action' and acquired 3D models.

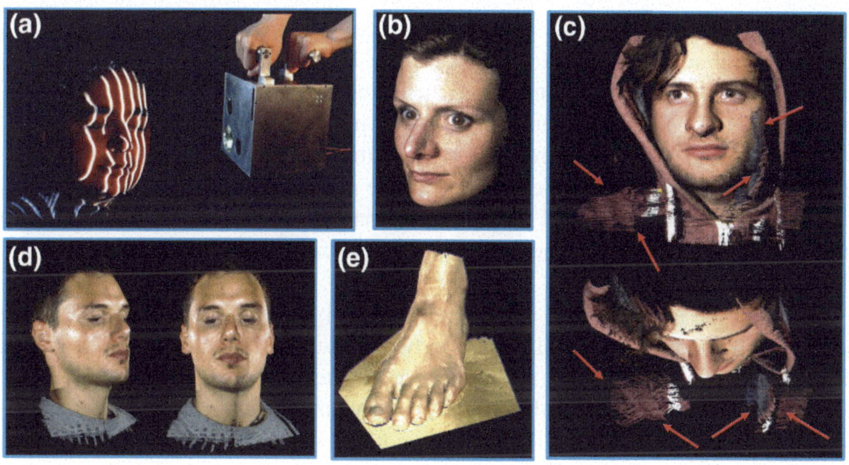

Fig. 4.9 Flying Triangulation face scanner. **a** Handheld scanner device during measurement. **b** 3D model from a frontal face measurement. **c** 3D model with outliers (red arrows). **d** 3D model from an ear-to-ear measurement, shown from two different directions. **e** 3D model of a foot also measured with the face scanner

[2]In order to work correctly, the Flying Triangulation registration algorithm is - as most other registration algorithms - limited to surfaces with sufficient 3D structure. For example, the single 3D views of plain walls or perfect spheres cannot uniquely be registered to each other, since there are too many 'valid' possibilities for the alignment. See [4, 9] for more information.

As Fig. 4.9c reveals the sensor also suffers from outliers caused by a too short measurement range. 100 mm might be sufficient for a frontal face measurement as shown in Fig. 4.9b. For a full measurement of the human head, as shown in Fig. 4.9d, inexperienced users will leave the measurement volume during the scanning process.

The face scanner comprises a second (color-) camera which is able to acquire the surface color texture during an additional flash-exposure. As mentioned, this additional camera offers a new possibility for outlier correction. The method will be discussed in the next chapter. Detailed information about the Flying Triangulation face scanner can be found in [10, 11].

Body Scanner (2011)

The Flying Triangulation body scanner represents a further rescaling of the measurement principle. The scanner is tailored for the measurement of large objects like human bodies, sculptures or car bodies and is also able to acquire whole scenes like furniture arrangements, archeological excavations or forensic relevant crime scenes.

Beside the need to measure large objects, one of the main purposes for the development of the body scanner was to get rid of the indexing problem caused by an insufficient measurement depth. The body scanner solves this problem to a certain extent by scaling the Flying Triangulation principle up to a comfortable measurement volume of $800 \times 800 \times 550\ \text{mm}^3$ at a stand-off distance of 1000 mm. By this means, the measurement process is considerably simplified even for inexperienced users. At the time of development white light LEDs could not produce sufficient luminance for the large measurement volume. Therefore lasers in combination with DOEs (diffractive optical elements) are used for the projection. Due to the increased speckle noise, the body scanner has a relatively large measurement uncertainty of $\delta z \leq 1.1$ mm in the whole measurement volume.[3]However, applying lasers also has advantages: First, the projector design is streamlined and simplified. No Köhler illumination or elaborate cooling is required. Second, due to the monochrome light source a narrow band color filter can be applied in front of the main camera, so only the line signals are visible. This enables a 3D measurement in even extreme ambient light conditions, e.g. in sunlight.

Although the measurement volume is sufficiently large for a comfortable measurement, the body scanner still suffers problems related to the indexing problem. Due to the large depth of field (small aperture) of the laser projection, the projected lines are still sharp far outside the measurement volume. Hence, objects (e.g. walls) behind the measured object will also cause outliers in the 3D data set. An example was already shown in Fig. 4.4b. Another problem to be mentioned is even *caused* by the large measurement volume: The sensor is just an up-scaled version of the dental or face scanner. This means the *3D data resolution* $1/\delta x_{dat}$ is still restricted by Eq. (4.4), and yields a very small value for this sensor.[4] An object with a lateral

[3]Evaluated on plain white paper. On other surfaces like human skin, the noise may differ due to volume scattering effects.

[4]Note that the *data density* has comparable values for all sensors, since it is not dependent from the field width.

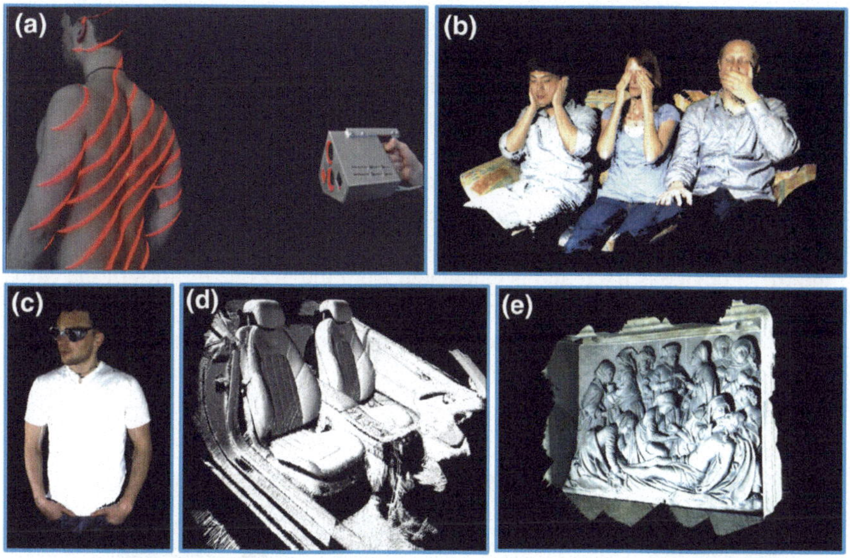

Fig. 4.10 Flying Triangulation body scanner. **a** Handheld scanner device during measurement. **b**, **c** 3D models from the measurement of persons. **d** 3D model of a car cabin. **e** 3D model of a stone relief

extension considerably smaller than the measurement field (at least in one dimension) will not 'host' sufficient crossing points for a robust registration on its surface. Of course, this is the case for all Flying Triangulation sensors. For the body sensor (11 resp. 10 projected lines) this is of specific relevance, since that scenario occurs frequently, when a standing human person is measured from the side.

Figure 4.10 shows a hand-guided measurement with the body scanner and some related 3D models. More information about the body scanner can be found in [12, 13].

MEG Sensor (2013)

The Flying Triangulation MEG sensor is not a 3D *scanner* in the common sense. It is more a *sensor* for the three-dimensional detection of movements developed for a specific medical application. The purpose of the sensor can be simplified as follows: Patients with brain diseases, like epilepsy, are commonly examined with a so-called magnetoencephalograph (MEG). This device for functional neuroimaging allows for a detection of active brain regions by recording magnetic fields over several minutes. The goal is to localize brain regions involved in an epileptic seizure to remove them surgically.

The MEG acquires data only inside its own coordinate system. The position of the patient's head during the measurement is unknown. For a precise detection of diseased brain regions inside the 'head coordinate system' the relative position between head and MEG device has to be evaluated.[5] Existing methods use small coils attached on salient points in the face for co-registration. Since activated coils spoil the MEG measurement drastically, a position detection can be only performed before and after the measurement. The head motion *during* the MEG measurement remains unknown.

With the Flying Triangulation MEG scanner the head motion detection can be performed *during the whole* MEG measurement. During the measurement the sensor is running and acquires sparse 3D line profiles of the face. The sensor position relative to the MEG device is fixed and is calculated from the 3D line profiles of the MEG device, which are acquired in the same image (see Fig. 4.11). After the measurement only the sparse face data are registered and the 'virtual sensor path' is calculated. The inverse of the virtual sensor path gives exactly the desired head motion curve.

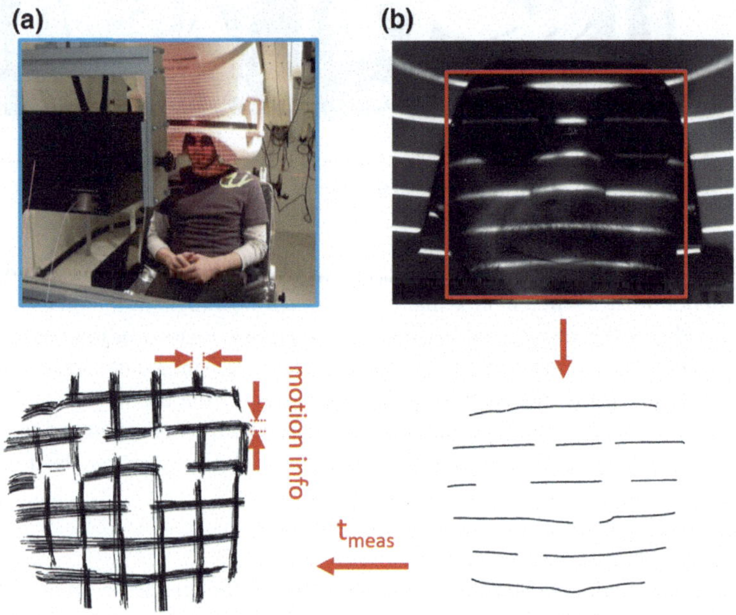

Fig. 4.11 Flying Triangulation MEG sensor for the three-dimensional detection of head motions during an MEG measurement. **a** Running measurement: Patient under the MEG device, placed in front of the Flying Triangulation sensor. **b** Evaluation of the head motion: From each single camera frame, a sparse 3D model is calculated. After the total measurement time t_{meas} and the registration of all sparse 3D models, the head motion during the measurement can be calculated via the 'inverse virtual sensor path'

[5]Commonly, the detection of affected brain regions with an MEG is supplemented by an electroencephalograph (EEG), which is able to measure electric fields in the brain by placing a hood of electrodes directly on the patients head. For correct localization the position of the electrodes also has to be co-registered with the head's surface, especially with the region of the face [16–18]. The relating 3D measurements can be performed with the already introduced Flying Triangulation face scanner.

Figure 4.11 shows the sensor in action inside the shielded MEG chamber as well as sparse line profiles, which can be used for the registration and path evaluation. In order not to spoil the MEG measurement, electric parts, inducing magnetic fields had to be avoided. The projection uses fiber coupled lasers, where all electronic parts are placed outside the shielded chamber. The used camera inside the chamber is shielded with a special housing. The sensor acquires data at 30 Hz along 11 resp. 9 lines within a measurement volume of $200 \times 150 \times 180$ mm^3. According to [19], the achieved measurement uncertainty is below 1.4 mm.

Although the Flying Triangulation MEG sensor is not applied for free hand-guided scanning and used in a known environment, it nevertheless suffers the indexing problem: For high precision motion tracking an *accurate registration* is necessary. The registration accuracy improves with the number of projected lines. As shown in Fig. 4.11 only about 7 resp. 6 lines are used for the detection of the head motion, whereas the rest is required for position detection of the MEG device. More lines would *significantly improve* the result, but they cannot be projected due to the restrictions of Eq. (4.4). More information about the Flying Triangulation MEG sensor can be found in [19, 20].

MIDI Scanner (2013)

The Flying Triangulation MIDI scanner is focused on applications in virtual reality. Experience has shown, that especially in this regime a scanner with a measurement volume between face and body scanner is advantageous for most applications.

For high user friendliness the scanner actively tackles the above described problems by a larger number of projected lines (25 vertical resp. 23 horizontal) and an extended measurement depth of 180 mm (measurement volume: $400 \times 300 \times 180$ mm^3). As the basic principle of Flying Triangulation is still trapped within the limits of Eq. (4.4), the user friendliness has its price: An extremely small triangulation angle of $\theta \approx 2°$ causes a low precision not better than $\delta z \approx$1.5–2 mm (see Fig. 4.12b). For the desired virtual reality application this can be concealed by post processing methods like smoothing or high resolution texture mapping. For metrology tasks the scanners field of application is very restricted, though.

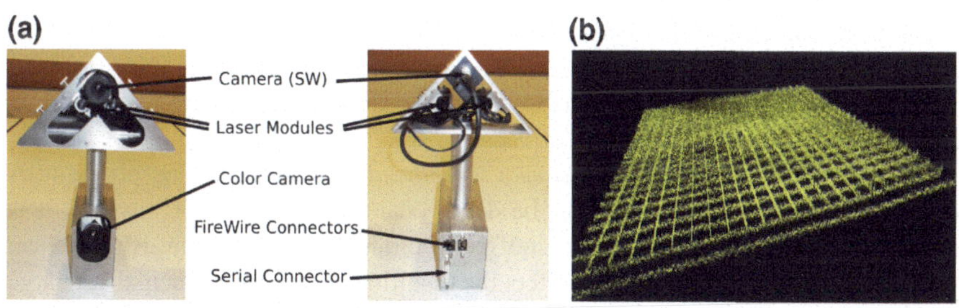

Fig. 4.12 **a** MIDI scanner with description of components. **b** Two sparse 3D models (horizontal and vertical) from the measurement of a planar surface indicating the high noise level of the MIDI scanner

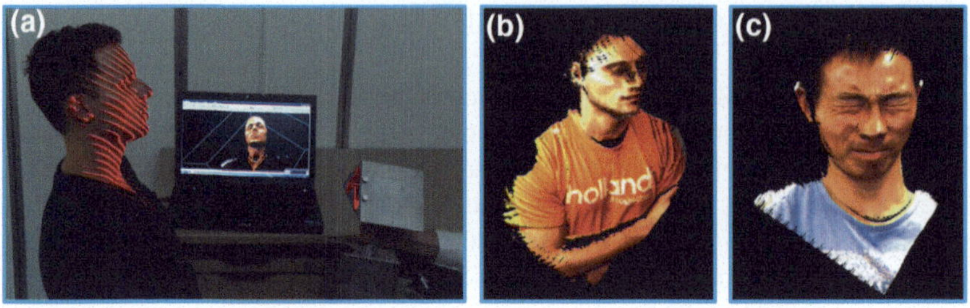

Fig. 4.13 Flying Triangulation MIDI scanner. **a** Measurement of a person with real-time preview of the measured scene in the background. The high number (compared to other Flying Triangulation scanners) of projected lines is visible. **b** and **c** 3D models acquired with the MIDI scanner

The MIDI scanner also uses laser modules with attached DOEs for the projection and a color filter to enable outdoor measurements. The second color camera is mounted under the sensor handle (see Fig. 4.12a) and helps to balance the sensor during the measurement.

Due to the increased number of projected lines the MIDI sensor can be considered as a first approach towards a *'single-shot 3D movie camera'*. However, it is restricted to applications with low requirements of lateral resolution and depth precision (comparable to *Microsoft Kinect One*®).

Figure 4.13 shows a hand guided measurement with the MIDI scanner and some related 3D models. More information about the MIDI scanner can be found in [5, 21].

4.4 Conclusion

The last chapter indicated that multi-line triangulation is a powerful tool for optical 3D metrology. Due to its scalability, it enables a wide range of applications. With proper optical design multi-line triangulation is able to deliver *very precise* 3D data with *high data resolution* in the direction of the projected lines.

The last chapter also demonstrated that the data resolution perpendicular to the line direction has limits. The scope of possible multi-line triangulation sensors is trapped in the boundaries of Eq. (4.4) resulting in a tradeoff between *precision*, *measurement depth* and *3D data resolution*.

Flying Triangulation is designed as a *workaround* on the indexing problem. A dense point cloud does not have to be acquired in *one* single shot. Instead, it is composed from *many* sparse single-shot point clouds by registration. Rigid object movements are allowed and do not affect the resulting 3D model. Principally, a large measurement range paired with high precision can be reached by projecting only very few lines. This would slightly increase the scanning time, which is negligible for most applications. However, in practice this is not possible, since the registration needs a certain number of crossing points to work robustly.

As a consequence, Flying Triangulation is *also* affected by the indexing problem, which is the reason that each of the discussed Flying Triangulation scanners displays at least one 'weak spot'.

The problem gets even worse, if the object movements become *non rigid*. Then, a scanning-based approach like Flying Triangulation, is not able to acquire a dense point cloud anymore. A new measurement principle is required, which is able to acquire 3D data along many projected lines within *one* single shot – the *'single-shot 3D movie camera'*. Within the limits of Eq. (4.4) such a sensor is *not possible*. For example, a scanner with the same field and triangulation angle as the face scanner, projecting 150 instead of 12 lines, would have a unique measurement depth smaller than 10 mm!

The next chapter introduces a new basic approach, which leads to several methods for the solution of the indexing problem. Eventually, these methods will lead to the *'single-shot 3D movie camera'*, but they can be applied to tackle the 'weak spots' of Flying Triangulation as well.

References

1. S. Ettl, O. Arold, P. Vogt, O. Hybl, Z. Yang, W. Xie, G. Häusler, Flying triangulation: a motion-robust optical 3d sensor principle, in *Fringe* (Springer, Heidelberg, 2009), pp. 1–4
2. S. Ettl, O. Arold, Z. Yang, G. Häusler, Flying triangulation - an optical 3D sensor for the motion-robust acquisition of complex objects. Appl. Opt. **51**(2), 281–289 (2012)
3. F. Willomitzer, S. Ettl, O. Arold, G. Häusler, Flying triangulation - a motion-robust optical 3D sensor for the real-time shape acquisition of complex objects. in *Proceedings of the 3rd International Topical Meeting on Optical Sensing and Artificial Vision, AIP 1537* (2013), p. 19
4. O. Arold, S. Ettl, F. Willomitzer, G. Häusler, Hand-guided 3D surface acquisition by combining simple light sectioning with real-time algorithms (2014), arXiv:1401.1946
5. S. Ettl, Introductory review on 'Flying Triangulation': a motion-robust optical 3D measurement principle. in *Contemporary Physics* (2014)
6. C. Matabosch, 3D hand-held sensor for large surface registration. Dissertation, University of Girona (2007)
7. C. Matabosch, D. Fofi, J. Salvi, E. Batlle, Registration of surfaces minimizing error propagation for a one-shot multi-slit hand-held scanner. Pattern Recognit. **41**, 2055–2067 (2008)
8. F. Willomitzer, S. Ettl, C. Faber, G. Häusler, Single-shot three-dimensional sensing with improved data density. Appl. Opt. **54**(3), 408–417 (2015)
9. O. Arold, Flying Triangulation - handgeführte optische 3D Messung in Echtzeit. Dissertation, University Erlangen-Nuremberg (2013)
10. F. Willomitzer, Z. Yang, O. Arold, S. Ettl, G. Häusler, 3D face scanning with "Flying Triangulation." in *Proceedings of the 111th DGaO Conference* (2010), p. 18
11. F. Willomitzer, FlyFace – 3D-Gesichtsvermessung mittels "Flying Triangulation". Diploma Thesis, University Erlangen-Nuremberg (2010)
12. F. Huber, O. Arold, F. Willomitzer, S. Ettl, G. Häusler, 3D body scanning with "Flying Triangulation". in *Proceedings of the 112th DGaO Conference* (2011), p. 18
13. F. Huber, Bewegungsrobuste 3D-Vermessung großer Objekte mittels Flying Triangulation. Diploma Thesis, University Erlangen-Nuremberg (2011)
14. Z. Yang, Ein miniaturisierter bewegungsunempfindlicher 3D-Sensor. Master Thesis, University Erlangen-Nuremberg (2009)

15. Z. Yang, W. Xie, O. Arold, O. Hybl, P. Vogt, K. Bobey, G. Häusler, A 3D sensor for intraoral metrology. in *Proceedings of the 110th DGaO Conference* (2009), p. 19
16. S.S. Dalal, S. Rampp, F. Willomitzer, S. Ettl, Consequences of EEG electrode position error on ultimate beamformer source reconstruction performance. Front. Neurosci. **8**(42), 1–7 (2014)
17. S. Fouladi-Movahed, Novel method to localize EEG electrodes employing the optical 3D sensor "Flying Triangulation". Master Thesis, University Erlangen-Nuremberg (2012)
18. S. Ettl, S. Rampp, S. Fouladi, S.S. Dalal, F. Willomitzer, O. Arold, H. Stefan, G. Häusler, Improved EEG source localization employing 3D sensing by "Flying Triangulation", SPIE optical metrology, in *International Society for Optics and Photonics* (2013)
19. A. Sadeghzadeh, Head tracking during magneto-encephalography (MEG) data recording via "Flying Triangulation". Master Thesis, University Erlangen-Nuremberg (2013)
20. S. Ettl, A. Sadeghzadeh, F. Willomitzer, O. Arold, S. Rampp, H. Stefan, G. Häusler, Management of head motion during MEG recordings with Flying Triangulation. in *Proceedings of the 114th DGaO Conference, A13* (2013)
21. B. Littau, Flying Triangulation "MIDI"-Sensor. Internship Report, University Erlangen-Nuremberg (2013)

Chapter 5
Solving the Problem with an Additional Source of Information

One fact became more than obvious from the discussions of the last chapters:

> *A triangulation-based single-shot principle with nearly* 100% *density of uncorrelated 3D data, also resolving high object frequencies, is* ***impossible!***

Line triangulation already displays important features of a 'perfect' single-shot principle but lacks high data density. Hence, current line triangulation sensors, like Flying Triangulation, are far away from being 'perfect'. The reason is the profound *ambiguity problem*.

The majority of measurement principles (see Chap. 3) buys correlation information by sacrificing feature and data resolution or/and temporal resolution.

> *What can be done to preserve the big advantage*[1] *of single-shot principles without sacrificing too much data density? Which additional modality can be exploited to obtain correspondence information?*

Between the discussed modalities, such as *color* or *polarization*, one modality turned out to be very promising: ***Additional perspective information*** provided by ***additional synchronized cameras***.

A sensor, which comprises a multitude of synchronized cameras, each taking a picture at the same time, is (according to the definition in this thesis) still *'single-shot'*, since the temporal advantage of single-shot principles is preserved.

The following chapter explains the basic idea of exploiting more than one camera to obtain correspondence information for multi-line triangulation. The two (linked) basic questions to be answered are:

---

[1] To memorize: The big advantage of single-shot principles is that the exposure time for a 3D height map is *uncoupled* from the camera frame rate. Assuming flash exposure with sufficient light, the exposure time for a 3D height map can be principally very short, even if the time for a single camera frame is much larger.

F. Willomitzer, *Single-Shot 3D Sensing Close to Physical Limits and Information Limits*, Springer Theses,
https://doi.org/10.1007/978-3-030-10904-2_5

- How can the ambiguity problem in Flying Triangulation be solved? The ambiguity problem provokes outliers which lead to false registration results.
- How to increase the data density, without sacrificing precision or measurement range? A solution of this problem would finally enable the desired *'single-shot 3D movie camera'*.

It is emphasized that all principles discussed in this chapter are ***not*** related to classical active or passive stereo, even if more than one camera is applied! (Single-shot) stereo sensing *exploits neighborhood information to a large extent*. This is explicitly ***avoided*** in the following to *preserve high feature resolution*.

5.1 Outlier Detection with Additional Perspective Information

The formation of false 3D data (outliers) in multi-line triangulation, caused by indexing ambiguities, was explained in Sect. 4.2: A part of the object surface is illuminated by the lines outside the unique measurement depth ΔZ. The signal is imaged in the false region of uniqueness on the camera chip, where a false line index is assigned. This process is again shown in Fig. 5.1a. The outlier cannot be distinguished from the correct data point, since it is *located on the same ray of sight* of the camera.

Now, the additional perspective information provided by a second camera (from now on: camera C_2) comes into play. As shown in Fig. 5.1b, the resulting geometry allows a clear separation of outliers and correct 3D data: If the 3D data evaluation is *correct*, the generated data point in space (black circle) is seen by C_2 under the same ray of sight (black dashed line) as the signal from the object surface (pink dot with pink dashed line). The (falsely) evaluated outlier does not satisfy this condition!

This information is exploited by the so-called *'signal-back-projection-approach'*, which allows for a *detection*, and even *correction* of outliers in line triangulation.

5.1.1 The Signal-Back-Projection-Approach for Flying Triangulation

The *signal-back-projection-approach* was mainly developed to get rid of indexing ambiguities in Flying Triangulation. In most Flying Triangulation setups, a second camera for the acquisition of color texture via additional (white) flash exposure is already implemented. This means that the additional perspective information provided by this camera is 'for free'.[2] Further refinements of this approach in the following sections will finally enable the *'single-shot 3D movie camera'*.

[2]Of course, only from a hardware-based point of view the information is 'for free'. From the information-theoretical point it is not, as a second camera picture is acquired.

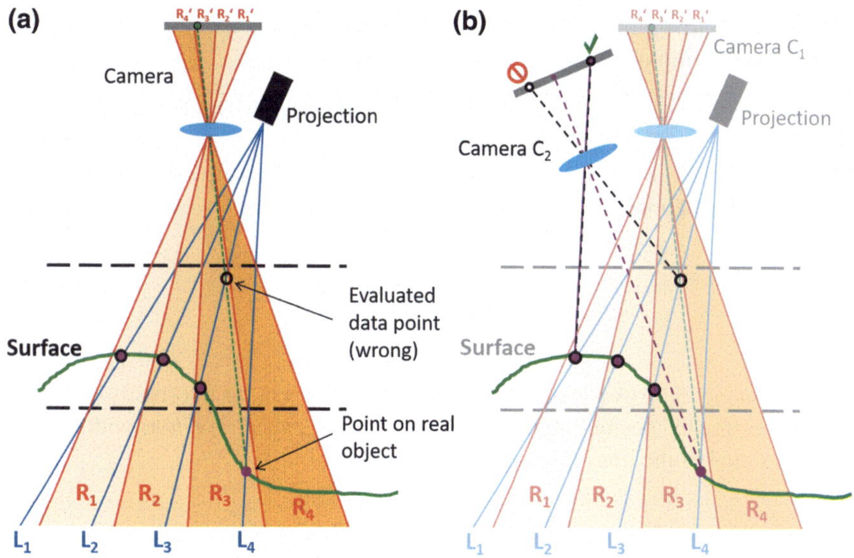

Fig. 5.1 Outlier detection with a second camera. **a** Outlier and correct data point are not separable, if only one camera is used. **b** Outlier detection by a second camera, which provides additional perspective information

How to quantify and exploit the additional perspective information, provided by C_2, for correct line indexing? A quite intuitive approach would be the following: Evaluate 3D points for both cameras and compare the coordinates of all points in virtual 3D space. If a point, originating from C_1, shares its position with a point, originating from C_2, it is expected to be *correct*. Such an approach would principally work, but it is technically elaborate: First, an additional longitudinal 'z-calibration' (see Sect. 8.1) would be required for C_2 (in Flying Triangulation setups, the color camera is only laterally 'xy-calibrated' by default [1, 2]). Second, due to the different position and triangulation angle of C_2, the unique measurement range of C_2 would have a completely different length and the measurement volumes would be tilted against each other. Last, the comparison of large point clouds in 3D space is computationally intensive.

A more simple procedure is the proposed *'signal-back-projection-approach'* [3–5]. It does *not rely on additional calibration* of C_2 and is *not restricted to unique measurement ranges*. Moreover, it enables a comparison of points in *2D-space*, which is *less complex*.

The approach is based on the back projection of the evaluated (virtual) 3D points onto the chip of the second camera. The basic work flow is as follows:

- Evaluation of 3D data with C_1. Like shown in Fig. 5.1a, the data could be indexed falsely.
- Back projection of each evaluated 3D point from the virtual 3D space onto the chip of C_2 (see Fig. 5.1b).

- Direct comparison of the back projection position with the observed intensity signal on the chip of C_2.

Figure 5.1b illustrates the procedure for a correctly indexed 3D point and an (incorrectly indexed) outlier: Each evaluated data point (x_e, y_e, z_e) in space is directly assigned to a back projection coordinate $(x'_B, y'_B)_{C_2}$ on the chip of C_2 via the calibration transformation K_{C_2}:

$$\begin{pmatrix} x'_B \\ y'_B \end{pmatrix}_{C_2} = K_{C_2} \begin{pmatrix} x_e \\ y_e \\ z_e \end{pmatrix} \quad (5.1)$$

As shown in Fig. 5.1b by a green check mark, the constraint for a correctly evaluated 3D point is, that the position of its back projection $(x'_B, y'_B)_{C_2}$ *overlaps* with the real intensity signal of the lines $(x'_S, y'_S)_{C_2}$ onto the chip of C_2:

$$\begin{pmatrix} x'_B \\ y'_B \end{pmatrix}_{C_2} = \begin{pmatrix} x'_S \\ y'_S \end{pmatrix}_{C_2} \quad (5.2)$$

To demonstrate the basic work flow of the described outlier detection, the approach is applied to the already shown outlier-example of Fig. 4.4b: A person is placed in front of a wall, whereas the wall is outside the unique measurement range ΔZ of the used sensor (here: body scanner). The work flow for one single-shot 3D view is shown in Fig. 5.2: Each camera acquires an image of the object with on-projected

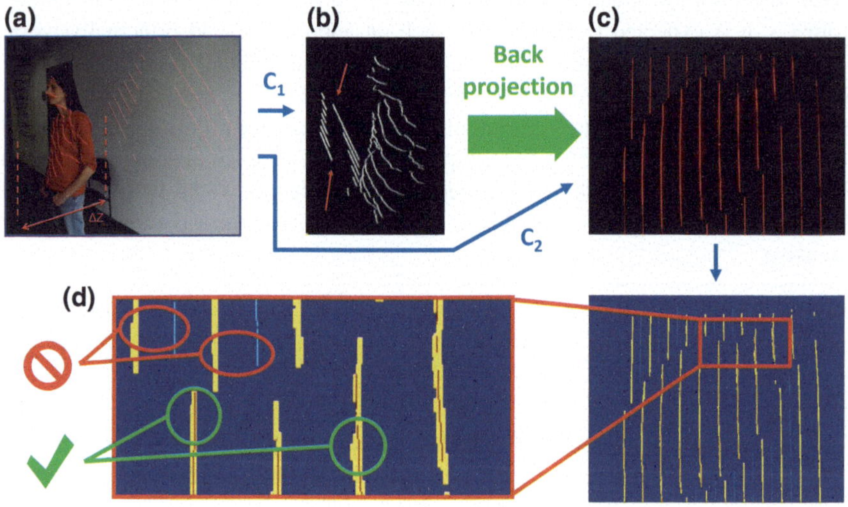

Fig. 5.2 Work flow of the signal-back-projection-approach. Both cameras acquire an image of the measured scene (**a**). The 3D view (**b**), which is directly calculated from the image of C_1, is back projected onto the image of C_2 (**c**). If the back projection in (**d**) overlaps with the intensity signal (red on yellow), the corresponding 3D point is correct

lines (Fig. 5.2a). A 3D view is directly calculated from C_1 (Fig. 5.2b). This 3D view contains outliers marked by red arrows. The 3D view is back projected onto the image of C_2 (Fig. 5.2c), which results in Fig. 5.2d. Here, the yellow regions mark the positions of the line intensity maxima in C_2, widened by a certain 'search region'. The thin red and blue lines mark the back projected 3D data. If a back projected point is located in a yellow region, it is considered to be *correct* (red on yellow), if not (blue and yellow separated), the point requires *further processing*.

A first (and not bad) approach for positive-checked outliers is to simply delete them. The sensor acquires a multitude of points in a very short time. Under the assumption, that the object to be measured is placed *inside* the unique measurement range for most of the acquisition time, deleting emerging outliers is surely a proper procedure. Moreover, objects behind ΔZ, like the wall in the shown example, are not of interest in many cases.

Things change if the object structures to be measured are *deeper* than ΔZ, or in simple terms: if the *unique measurement range is too short* for the measured object. Then, distant and close surface parts cannot be measured at the same time.

What about moving the sensor in z-direction for a subsequent scan of different depths? For Flying Triangulation this is only possible if all surface parts are *sufficiently connected*, in order to guarantee for a working registration. Wall and person in Fig. 5.2a could not be measured together, if the outliers are simply deleted.

Hence, a method that *corrects* the outliers (instead of simply deleting them) is necessary. In fact, such a method would *expand* the effective (usable) measurement range ΔZ_e to a value significantly larger than ΔZ.

The proposed solution consists of an iterative trial-and-error method. The scheme is displayed in Fig. 5.3: A new line index ($n \rightarrow n \pm 1$) is assigned to each 3D point that does not pass the back projection (BP) check. The point is evaluated again with the new line index and checked again. This is done iteratively, until the point passes the back projection check.

For the shown example, the signal-back-projection-approach is applied to all 120 sparse 3D views that were subsequently acquired from the scene while slightly moving the sensor. From roughly 1 Mio acquired 3D points, 250,000 are identified as outliers. The false points are *deleted* or *corrected* and the processed 3D views are registered to a full 3D model. The results are shown in Fig. 5.4. Figure 5.4b shows the original 3D model with outliers (red arrows) solely evaluated from C_1. The results for *removal* and *correction* of outliers are shown in Fig. 5.4c, d. No

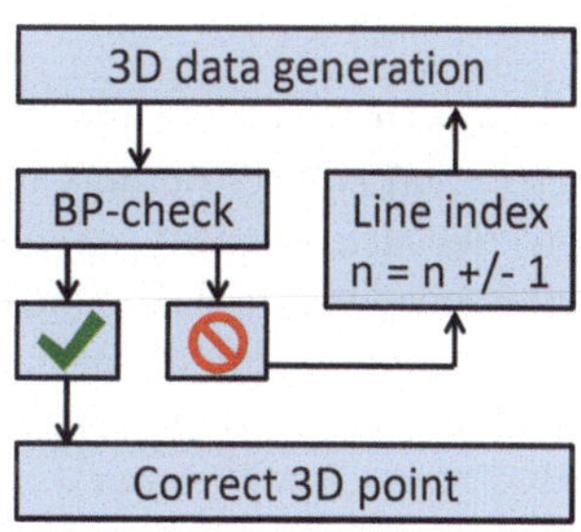

Fig. 5.3 Scheme for the correction of outliers, based on an iterative trial and error increase and decrease of the line index, until the point passes the back projection check

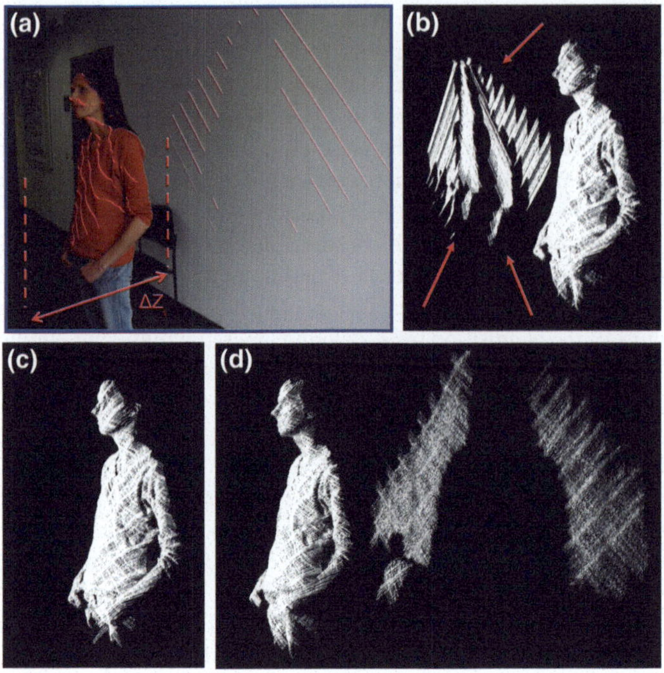

Fig. 5.4 Results of the signal-back projection-approach. **a** Measured scene. **b** Evaluated 3D model without signal-back-projection. Approximately 1/4 of the acquired points are detected as outliers. **c** Identified outliers deleted. **d** Identified outliers corrected. No remaining outliers are visible in the processed datasets

remaining outliers are visible in the 3D data sets. In Fig. 5.4d, the 3D data of the wall are correctly indexed and correctly located (no 3D data in the occlusion area). The 3D model now contains correct data from the *full* scene of Fig. 5.4a, although the unique measurement range ΔZ, defined by Eq. (4.4), is considerably smaller ($\sim 1/3$ of the depth of the full scene).

5.1.2 Potentials and Limits of the Signal-Back-Projection-Approach

As shown in the last section, the signal-back-projection-approach is able to effectively *detect* and *correct* outliers in Flying Triangulation, which results in an *increased effective measurement range* $\Delta Z_e > \Delta Z$. For the first time, the sensor principle is not trapped within the boundaries of Eq. (4.4) anymore!

Furthermore, the signal-back-projection-approach *preserves all valuable features* of Flying Triangulation: First, the modified measurement principle is *still single-*

shot. The signal-back-projection-approach can be performed separately for each single camera frame, *independent from previous or subsequent exposures*. Since the computational effort is low, the method can be performed in real-time during the Flying Triangulation measurement (see below). Second, the method is performed *point-wise*. No space-intensive correlation operations[3] on the chip are required. The *high feature resolution* of Flying Triangulation is *preserved*.

Where are the limits of the proposed method, i.e. when does the method fail?

Theoretically, an outlier passes the back projection check if its back projection onto the chip of C_2 lands exactly on the image of another line. This case is demonstrated in Fig. 5.5. If the *positions of the back projected outliers* onto the chip of C_2 are assumed to be *statistically distributed*, a probability P_f that the back projection check of an outlier *fails*, can be calculated as

$$P_f = \frac{\Delta s'}{\Delta X'} \cdot (N_L - 1) \, . \tag{5.3}$$

$\Delta s'$ is the width of the defined 'search region' on the chip of C_2, represented in Fig. 5.2d by the width of the yellow lines.

Assuming (in agreement with the experiment demonstrated in Fig. 5.2) a search region of $\Delta s' \approx 5\ pix$, a chip width $\Delta X' = 1024\ pix$ and $N_L = 11$ projected lines (body scanner), the fail-probability reaches a value of $P_f \approx 5\%$. This is a quite good value for the example of Flying Triangulation ($N_L \approx 10$). Nevertheless, in setups

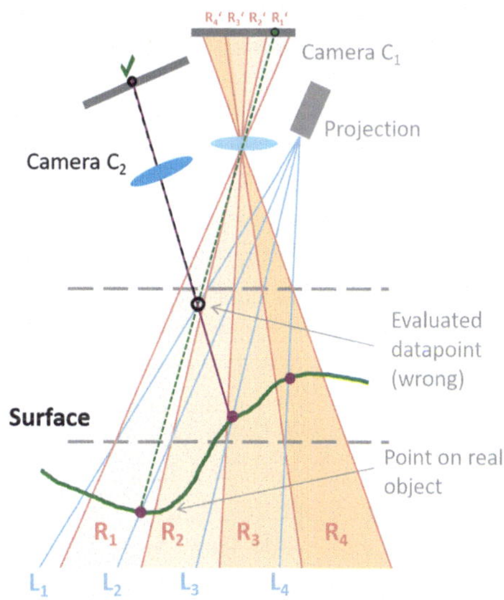

Fig. 5.5 Point-outlier constellation for a failing outlier-check

[3]Note, that this is the reason why the back-projection-approach is *not related to active stereo approaches*.

with higher line numbers ($N_L \approx 50$–100), P_f reaches a value which is no longer negligible. In the next sections improvements are discussed, which also enable the correct indexing of higher line numbers.

In the framework of the master thesis [6], the introduced signal-back-projection-approach was implemented in the Flying Triangulation evaluation software. Evaluated 3D points of sparse single-shot views are checked before they undergo the registration. Detected outliers can be deleted or corrected. The back-projection-approach seamlessly fits into the entire evaluation chain, which can still be performed in real-time with visual feedback of the corrected data on the screen.

By applying the signal-back-projection-approach, the effective measurement depth of the related sensors was increased *without any hardware changes* to $\Delta Z_e \approx 3 \cdot \Delta Z$. This is already at the limit of the depth of field for most Flying Triangulation sensors.

5.1.3 Possible Improvements, Part 1

If the positions of back projected outliers on the chip of C_2 are further on considered as statistically distributed, one method to improve the principle, i.e. reduce P_f, becomes very evident: ***more cameras!***

If, e.g., three cameras would be used, an outlier, created by C_1, would only pass the outlier check if it would be simultaneously back projected onto a line signal on the chips of C_2 and C_3.

For $N_{cam} \geq 2$ applied cameras the 'fail probability' P_f is reduced to[4]

$$P_f = \left(\frac{\Delta s'}{\Delta X'} \cdot (N_L - 1) \right)^{N_{cam}-1}, \tag{5.4}$$

Figure 5.6 demonstrates the influence of more than two cameras on the performance of the outlier detection. The object is a plaster bust. Only one single 3D view is considered. The 3D data are evaluated from the image of C_1, and processed via the signal-back-projection-approach together with the images of two more cameras, resulting in $N_{cam} = 3$. Data resolution, measurement range and triangulation angle of the 'main' sensor for the data generation (C_1 plus projection) is chosen in that way, that the sensor displays a *fivefold ambiguity*, meaning that Eq. (4.4) would deliver $\frac{N_L}{\Delta X} \cdot \Delta Z \cdot \tan\theta \approx \mathbf{5}$ instead of 1. The right-hand side of Fig. 5.6 displays the evaluation results for $N_{cam} = 1$ (no outlier check), $N_{cam} = 2$ and $N_{cam} = 3$ from a top view and a side view. Since each pixel of C_1 corresponds to 4–5 different line indices, the 3D model for $N_{cam} = 1$ displays a huge amount of outliers. It can be seen, that all outliers assemble to 3–4 additional 'phantom 3D models' in front and behind the true 3D model. The majority of outliers is removed after the back projection check with the second camera. Nevertheless, a considerable amount of

[4]Cameras with equal hardware (chip and optics) as well as perfectly overlapping fields assumed.

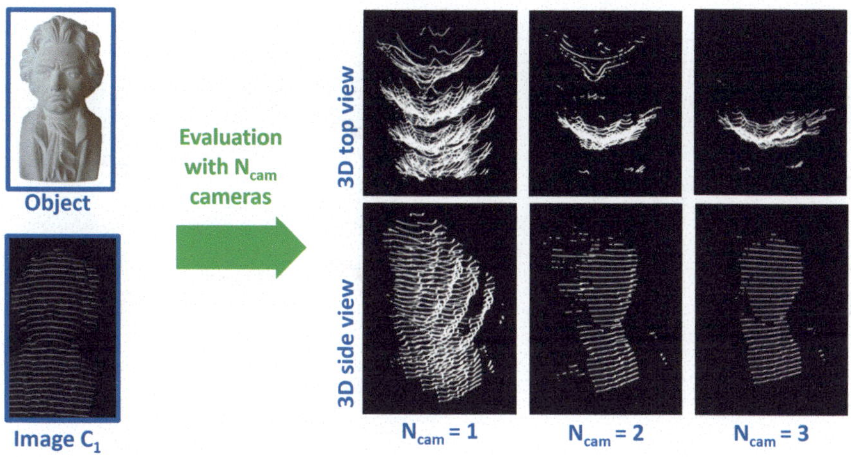

Fig. 5.6 Signal-back-projection with more than two cameras

outliers 'survives'. After the next back projection check with the third camera nearly all outliers are detected and removed. A third back projection check may detect and remove the rest.

The signal-back-projection-approach with $N_{cam} \geq 2$ cameras seems very promising even for ambiguity levels > 5. Nevertheless, even for Flying Triangulation sensors with a built-in color camera additional hardware would be required.

In the next sections, it will be discussed whether or not the method even has the potential to provide a *'single-shot 3D movie camera'*.

5.1.4 Side Note: The Tomographic Triangulation - Experiment

'Tomographic Triangulation' [7] is an experiment that takes the idea of applying more cameras for outlier suppression to the extreme. The question, that should be answered by the experiment, is:

> Is a *single-shot principle* with nearly 100% density of independent 3D data possible if a very large number of cameras is exploited?

The reader might guess the answer. In the following, a simulation[5] explains the basic idea (see Fig. 5.7), followed by real experiments. For the theoretical explanations a perfect Lambert-scattering surface is assumed.

In order to *provide a data point in each camera pixel without violating the sampling theorem*, a band limited sinusoidal pattern (as shown in Fig. 5.7a) is projected instead of a line pattern. The simulation is performed for $N_{cam} = 6$ cameras in a planar

[5]Figures related to the simulation were provided by Christian Faber, now at University of Applied Sciences, Landshut, Germany.

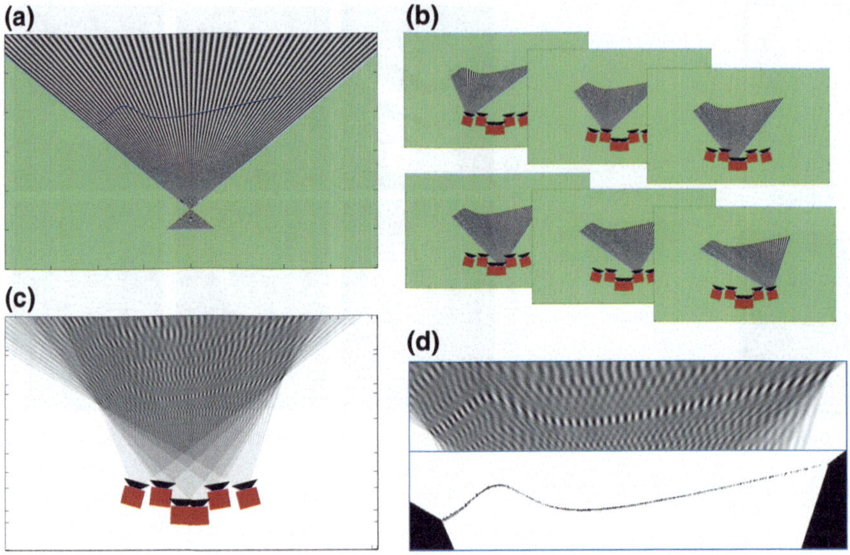

Fig. 5.7 Basic principle of Tomographic Triangulation: A sinusoidal pattern is projected onto the object surface (**a**) which is simultaneously observed with N_{cam} cameras (**b**). After back projection of all camera images in 3D space (**c**) and application of probability based Hough-concepts, the object surface is uniquely reconstructed (**d**)

geometry (x-z-plane). Each camera observes the object surface with the on-projected sinusoidal pattern from a different direction, meaning that each camera acquires a *slightly different image of gray values* (see Fig. 5.7b). For an evaluation of the object surface, each acquired camera image is *virtually back projected in 3D space* (in simulation: x-z-plane). Figure 5.7c, as well as the upper Fig. 5.7d display a superposition of all back projected camera images in 3D space. One can recognize, that the *contrast is maximal* at all coordinates of the object surface. In other words: For each point (x_i, y_i, z_i) of the object surface, the back projections of ***all*** N_{cam} cameras display the same gray value:[6]

$$I_1(x_i, y_i, z_i) = I_2(x_i, y_i, z_i) = \cdots = I_{N_{cam}}(x_i, y_i, z_i) \tag{5.5}$$

The basic idea of Tomographic Triangulation is therefore:

> Each *back projected gray value* from one camera chip *overlaps in 3D space* with $(N_{cam} - 1)$ *exactly equal gray values*, back projected from all other cameras, ***only*** at the *coordinates of the object surface.*

For the evaluation of the object surface, a *'Hough-concept'*, similar to the evaluation concepts of Computer Tomography is applied: The virtual 3D space is divided into voxels. In each voxel, the back projected intensity values from each camera are

[6]This is quite trivial, since the surface point (x_i, y_i, z_i) is the 'origin' of the gray value $I(x_i, y_i, z_i)$.

compared. If, in the ideal case, all back projected intensity values in a voxel are equal, the voxel is considered to be a part of the object surface. The bottom part of Fig. 5.7d displays all voxels (in black), where the standard deviation $\sigma_{N_{cam}}$ of back projected intensity values from all cameras is *zero*. The object surface is perfectly reconstructed without any ambiguities.

So much for the theory. In practice, electronic and coherent *noise* as well as *non cooperative object surfaces* make it impossible to reach a perfect equality ($\sigma_{N_{cam}} = 0$) of back projected intensity values in the surface-voxels. A threshold to $\sigma_{N_{cam}}$ has to be set, which gives rise to outliers. Figure 5.8a displays a simulated result in presence of noise (simulated $SNR = 10$) for $N_{cam} = 6$ cameras with a threshold of $\sigma_{N_{cam}} = 0.004$. The object surface is reconstructed, but some outliers remain. Theoretical considerations, as well as further simulations reveal, that the results in presence of noise can be significantly improved by projecting a *pattern with a high frequency* (close to the sampling limit) and by *applying many cameras* for the measurement. The number of outliers, induced by noise, *decreases exponentially* with N_{cam}, which means that even high noise levels could be handled with a sufficient number of cameras.

It is also important to note, that *Tomographic Triangulation is completely insensitive to surface texture* (varying surface reflectivity) and *background illumination*, since *only local intensity values* are compared.

A much more severe problem is the *angle dependent reflectivity* of surfaces which are not perfectly lambertian. Since each camera observes the surface from a different angle, each camera gets different intensity values for the same surface point of a non-lambertian surface. Figure 5.8b displays a simulation result with the same specifications as Fig. 5.8a ($SNR = 10$; $N_{cam} = 6$; $\sigma_{N_{cam}} = 0.004$), but with additional *angle dependent surface reflectivity*. The amount of outliers is larger and the surface is *not densely reconstructed*. This problem occurs especially at the zero crossings of the projected sinusoidal pattern and does not improve with more cameras. In practice, the only possible solution for this problem is the acquisition of an *additional 'white image'* (image taken with homogeneous white illumination) for *each* camera to compensate for the angle dependent surface reflectivity. This makes Tomographic Triangulation *not a single-shot principle* anymore!

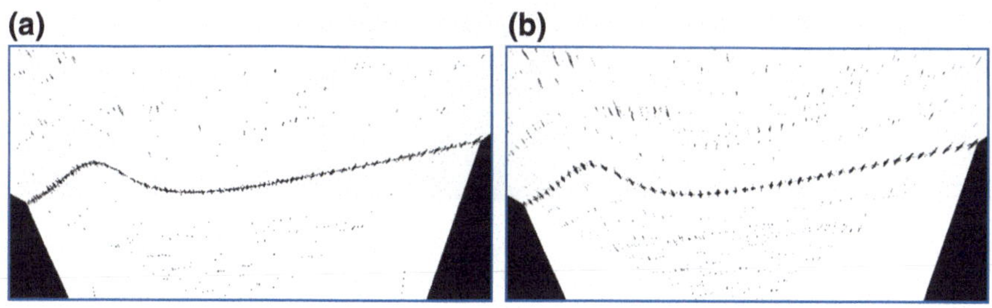

Fig. 5.8 Tomographic Triangulation in presence of noise and non-lambertian surfaces: Simulation results for $SNR = 10$ and threshold $\sigma_{N_{cam}} = 0.004$. **a** Perfect lambertian surface. **b** Non-lambertian surface

For the validation of simulated results, several experiments were performed. Figure 5.9b shows the measurement of a white plaster bust (Fig. 5.9a) with good lambertian surface properties. Figure 5.9d shows the measurement of a textured statue (Fig. 5.9c) with shiny surface parts. The measurement is performed with $N_{cam} = 15$ cameras, that are placed in a radius of $\sim$1000 mm around the object (see Fig. 5.10a). The measurement field is approximately 100×50 mm^2. It turned out that even for the plaster bust, a measurement without an additional 'white image' for each camera is *impossible*!

The 3D model of the plaster bust displays widely dense data, but still contains outliers. In the 3D model of the statue, outliers are also present. At very shiny object parts (e.g. the bow tie of the clown), the surface is not reconstructed densely (see red arrows in Fig. 5.10e), even *with* an additionally acquired white image. The data structure strongly resembles the simulation result of Fig. 5.8b.

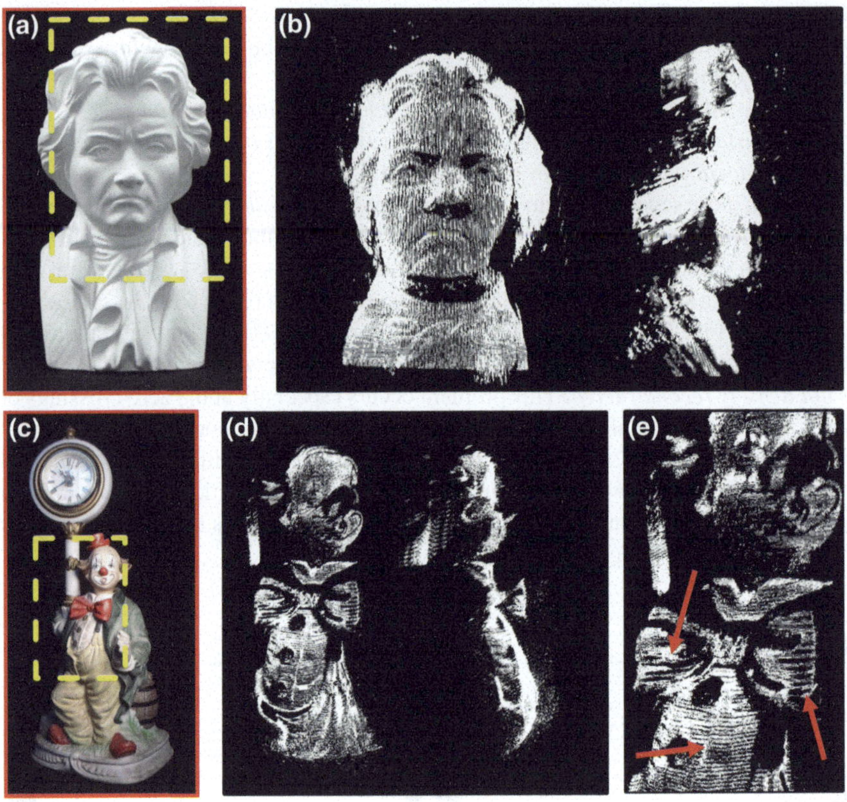

Fig. 5.9 3D models measured with Tomographic Triangulation. **a** and **c**: Measured objects with measurement field (yellow). **b** and **d**: Resulting 3D models from two perspectives. **e** Sparse surface reconstruction from 'non-lambertian' surface parts

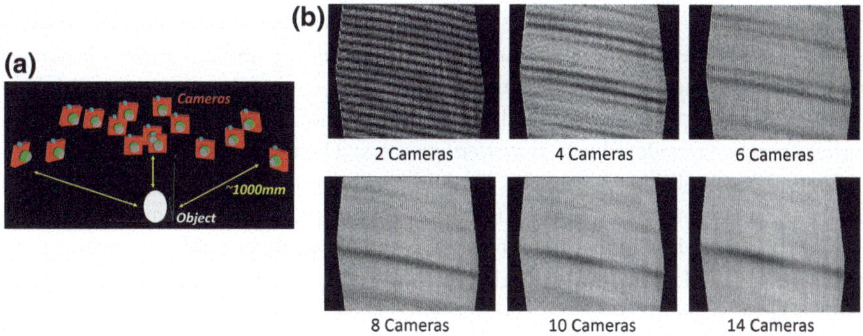

Fig. 5.10 **a** Positions of the cameras used for the evaluation (extracted from the calibration-data). **b** Influence of N_{cam} on the number of outliers

The experiment of Fig. 5.10b visualizes the influence of N_{cam} on the number of outliers. The object is a planar surface evaluated with different numbers of cameras. Each image of the figure shows a horizontal section through the measurement volume. The gray code displays the local variance $\sigma_{N_{cam}}$ in each voxel. Dark regions (low $\sigma_{N_{cam}}$) stand for a high probability that the surface is located there. As expected, the number of outliers is significantly reduced the more cameras are used for the evaluation.

The Tomographic Triangulation - experiment reveals again, that a 3D acquisition with 100% data density is impossible in single-shot (if no restrictions on the object surface are considered and color coding is not exploited). The principle suffers especially the *angle dependent reflectivity* at non-lambertian surfaces, that requires an additional 'white image' for each camera. If this image is acquired, the single-shot ability is lost.

The requirement of an additional 'white image' can be bypassed, if a binary pattern (i.e. again straight lines) is projected. However, this makes a 100% dense surface acquisition impossible. Nevertheless, the binary approach seems reasonable for the *'single-shot 3D movie camera'* if the density is considerably high.

Dense lines would also provoke many outliers, which can only be suppressed with a large amount of cameras. The next section reveals, that the amount of required cameras can be significantly reduced if several geometrical conditions of the setup are satisfied.

5.1.5 Possible Improvements, Part 2

In the last sections, a new approach for the detection and correction of outliers in line triangulation systems was introduced. The approach is based on the introduction of one or more additional cameras in the system. So far, the position of the cameras was more or less arbitrary in all discussed sensor setups: In the case of Flying

Triangulation, the new ideas were applied to an already finished setup, with a fixed geometry. In Tomographic Triangulation the position of all cameras is chosen as random as possible (see Fig. 5.10a).

Theory and experiment revealed, that -under these conditions- the formation of outliers can be considered as a *statistical process*: The more lines are projected, the higher is the probability for outliers. The more cameras are applied, the more outliers can be detected and suppressed. This probability decreases exponentially with the number of applied cameras. For Flying Triangulation, where only a few lines are projected per single-shot, the method is very effective, even with only two cameras. A sensor like the desired *'single-shot 3D movie camera'*, which should be able to project a very large line number per single-shot, would require many cameras. In practice, this would result in a bulky sensor setup, which requires extensive calibration and intensive computational effort for the evaluation of 3D data.

In this section, the influence of the *sensor geometry* on line indexing ambiguities is analyzed. As shown, the formation, detection and correction of outliers can be explained with a simple geometric model. The question is: *Can the amount of cameras be reduced by optimizing the geometry?*

The premature answer is: *Yes! ... under certain restrictions.*

The influence of geometry is investigated with a simple schematic model shown in Fig. 5.11: Lines are projected onto a surface, observed with a camera under the triangulation angle θ. For simplicity, the whole setup is telecentric and the surface is planar. In the example, the object should be measured within an effective measurement range ΔZ_e, which is *five* times larger than the unique measurement range ΔZ. Figure 5.11b displays the rays of sight, where C_1 observes the light signals, scattered from the object surface. Since the setup comprises *fivefold* ambiguity, each measured surface point (except the points on the edge of the field) has *five* possible coordinates: The true value (pink dot) and four outlier-values (black circle).

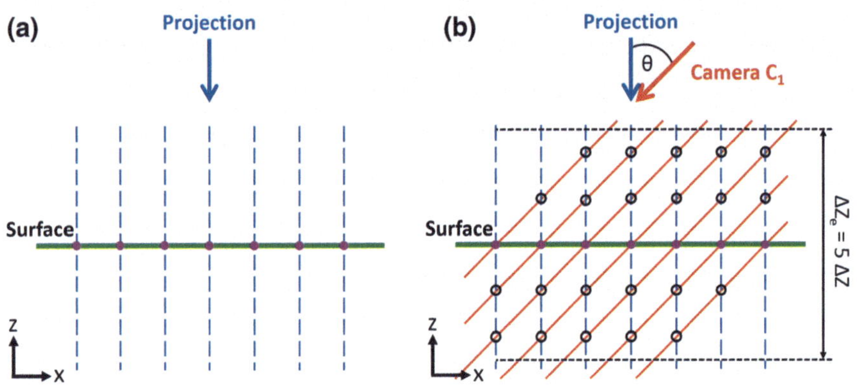

Fig. 5.11 Influence of camera geometry on outlier detection: initial setup. **a** Projection of lines onto a flat surface. **b** Observation with C_1 under the triangulation angle θ. The shown model comprises *fivefold* ambiguity ($\Delta Z_e = 5 \cdot \Delta Z$), meaning that each line signal on the camera chip leads to five possible 3D point-candidates

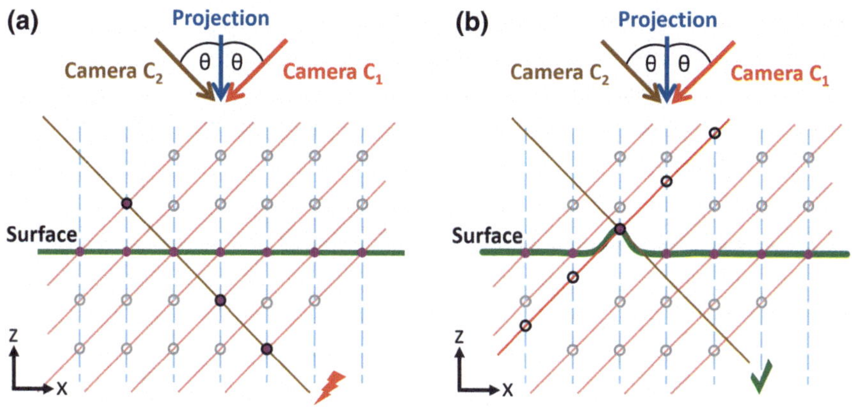

Fig. 5.12 Influence of camera geometry on outlier detection: symmetric arrangement. The outlier detection is dependent on the object surface

The question is: *Where* (i.e. under which angle) must a second camera be placed in order to *resolve the ambiguity* and detect all outliers? It becomes obvious from Fig. 5.12a, that the *symmetric* position with the same triangulation angle is very unfavorable in this case. However, this is only true for the special case of the planar, untilted surface. Approximately, this also yields for very smooth surfaces.[7] For surfaces containing higher spatial frequencies, the measured surface points with the related outliers might be distributed differently in the measurement volume. Then, even a symmetric setup can yield unique results as Fig. 5.12b displays.

How about asymmetric setups? As shown in Fig. 5.13a, correct indexing is here strongly dependent on the object surface again and cannot be guaranteed in the whole effective measurement range ΔZ_e. Obviously the statistical consideration of outlier formation is an appropriate model if no further assumptions about the object surface are made. This is important, since assumptions relating to the object or object surface restrict the general usability of the sensor.

Nevertheless, some assumptions can be made in good conscience. One of them is, that the *height variation of the object surface inside the measurement field is* ***always smaller*** *than the effective measurement range* ΔZ_e. For many sensors, ΔZ_e is limited by the depth of field (see Sect. 5.1.2) or is even *defined* by the maximal possible surface variations. In that case, a measurement outside ΔZ_e makes no sense.

If a maximal surface variation smaller than ΔZ_e, is assumed an angle θ_2 for the second camera can be defined, where ***all outliers can be detected***. In other words: The whole system produces **100%** ***correct data with only 2 cameras*** inside the ***whole effective measurement range*** ΔZ_e. Compared to conventional multi-line triangulation, this increases the measurement depth (*or* data density *or* precision - see Eq. (4.4)) in the

[7]Each projected line has a certain width and each back projection is evaluated in a certain search region (see Sect. 5.1.1) to compensate for noise and calibration errors, which means that outliers can also be located slightly beneath a ray of sight to be considered as correct points.

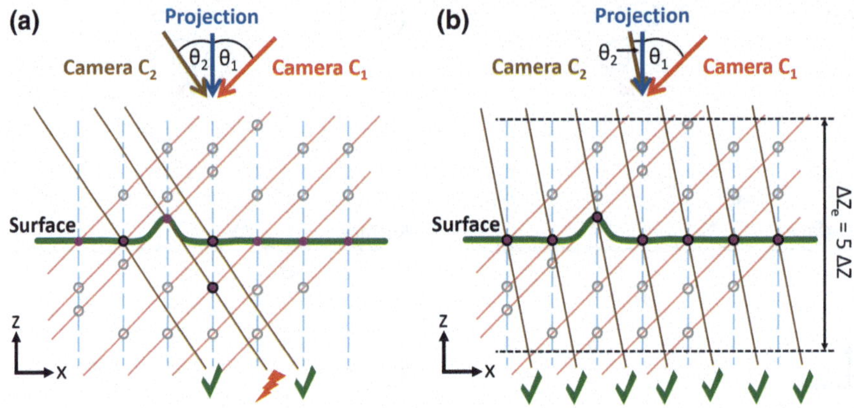

Fig. 5.13 Influence of camera geometry on outlier detection: asymmetric arrangement. **a** The outlier detection is still dependent on the object surface. **b** Solution: All outliers inside ΔZ_e are detected, if the second camera is placed under $\theta_2 << \theta_1$ in the setup

shown example by the factor of 5. As it will be demonstrated in Sect. 5.2.1, higher factors ~10 are possible in practice as well.

The basic idea is as follows (see Fig. 5.13b): An outlier can only be misinterpreted as correct point, if a ray of sight (brown line) from C_2 crosses *more than one* projected line (blue dashed line). Otherwise, an outlier will never be co-located with a true intensity signal from the surface at the same ray of sight.

Due to this reason it is advantageous to mount C_2 under an angle θ_2 which is ***considerably smaller*** than θ_1. As this method *guarantees* correct indexing in the whole effective measurement range, the outlier formation is *not statistical* anymore.

In the next section, further improvements and simplifications, based on the findings of this section are discussed, which finally lead to the *single-shot 3D movie camera.*

5.2 Towards the Single-Shot 3D Movie Camera

What has to be done to upgrade the discussed back projection idea towards the *single-shot 3D movie camera* – a sensor that provides a high density of 3D points, as well as a high precision, high measurement range and high feature resolution in *each single-shot*?

From the discussion in Sect. 5.1, two insights are of major importance for the further development: First, if the statistic-based *signal-back-projection-approach* is furthermore pursued, the desired high 3D data density in each single-shot would require a large amount of cameras to compensate for outliers. Second a single-shot principle, based on the projection and detection of *different gray values* is not practicable for the purposes of this thesis.

Hence, it seems straightforward to develop the *single-shot 3D movie camera* on the basis of *multi-line triangulation* - just as Flying Triangulation previously. For the desired high 3D data density, the correct evaluation of a ***multitude of lines*** is necessary. In order to 'save' cameras and computational resources, a variation of the signal-back-projection-approach, based on the *deterministic* arrangement of only two cameras was proposed in Sect. 5.1.5.

The formulated restriction, that *'a ray of sight from the second camera should not cross more than one projected line inside* ΔZ_e', can be expressed mathematically:

$$\tan\theta \leq \frac{\Delta X}{N_L \cdot \Delta Z_e} \tag{5.6}$$

This is the same equation, that limits precision, line number and unique measurement depth in ordinary multi-line triangulation systems (Eq. (4.4)) with the only difference, that ΔZ_e is much larger than ΔZ. In other words: If C_2, together with the projection, would form an *independent triangulation sensor*, this sensor must be *'unique' in the whole effective measurement range* ΔZ_e.

For the *signal-back-projection-approach* it is sufficient that C_2 is just used for a simple yes/no check of 3D points, meaning that it does not represent a *'full'* triangulation sensor. For the *single-shot 3D movie camera* the transformation of C_2 (plus projection) into a *'full'* triangulation sensor offers several benefits. Additional information is acquired, albeit additional calibration steps are necessary (see Sect. 8.1).

If this is done, the proposed line triangulation setup with two cameras can also be seen as *two independent measuring triangulation sensors that are coupled by the common used projection.* This is the basic aspect that enables the following *index-back-projection-approach.*

5.2.1 The Index-Back-Projection-Approach

The so-called *index-back-projection-approach* [8] enables the realization of the *single-shot 3D movie camera* (or short: *3D movie camera*) without the need for many cameras or complex algorithms. Compared to the discussed *signal-back-projection-approach*, the *index-back-projection-approach* can also provide *index-information* instead of performing a simple yes/no comparison.

This section introduces the *index-back-projection-approach* as well as a first embodiment of the *3D movie camera.* Moreover, first measurement results are presented. The potentials and limits of the method (e.g.: How many lines can be evaluated? Which precision can be reached?) as well as further improvements are discussed in the next chapters.

The *3D movie camera* requires three basic components:

- A **projector P** that projects a considerably large number of N_L narrow[8] lines, which are sharp in the whole effective measurement range.

[8] Within the limits of the sampling theorem.

- A **first camera $\mathbf{C_1}$** which encloses an angle θ_1 with the projection. θ_1 is chosen in a way, that the first triangulation sensor (P plus C_1) is *'unique'* in the whole effective measurement range ΔZ_e. This relation is expressed in Eq. (5.6). In other words: The unique measurement range ΔZ_1 of the first triangulation sensor includes the whole effective measurement range ($\Delta Z_1 \geq \Delta Z_e$).
- A **second camera $\mathbf{C_2}$** which encloses an angle θ_2 with the projection. θ_2 is chosen in a way, that the second triangulation sensor (P plus C_2) displays *high precision*. Due to Eq. (2.20), θ_2 should be considerably larger than θ_1, but small enough that the sensor does not suffer serious occlusion effects.

The fields of P, C_1 and C_2 should widely overlap. Note: The numbering of cameras has changed, compared to the last section (the camera with the small triangulation angle is now C_1). This is due to the chronological order of computational steps in the index-back-projection-approach (see below). A possible experimental setup is shown in Fig. 5.14. Figure 5.14a displays a front view of the setup. The lateral nodal point positions of projector P (white) and the two cameras C_1 and C_2 are marked by dashed lines. It can be seen that the small triangulation angle θ_1 could only be realized by placing C_1 *on top of* the projector. This would even be necessary if the projector would be less bulky - just because of the diameter of the front lenses. However,

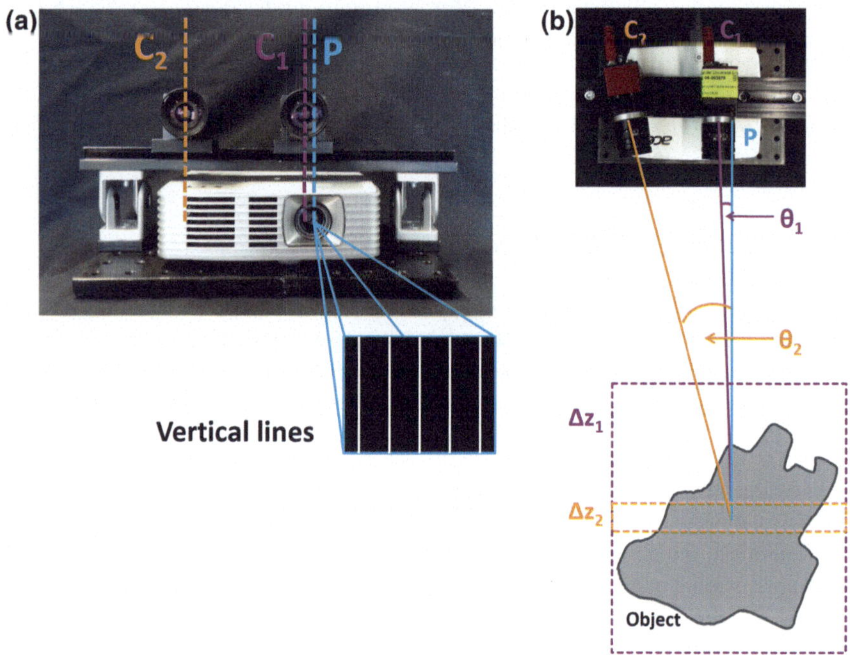

Fig. 5.14 Experimental setup, enabling the *index-back-projection-approach*. The setup comprises two independently measuring triangulation sensors, that are coupled by the projection. **a** Front view. **b** Top view, with schematic visualization of resulting angles and measurement volumes

this has no effect[9] on the measurement result, since only the *nodal point distance, perpendicular to the direction of projected lines* defines the triangulation angle. Figure 5.14b shows a top view of the setup and visualizes the different triangulation angles and unique measurement ranges.

For a single-shot 3D measurement, each camera simultaneously acquires an image of the object surface with the on-projected lines. If the two triangulation sensors would 'not know each other', each sensor for itself would be still trapped within the boundaries of Eq. (4.4): The data of sensor 1 are unique in the whole effective measurement range, but display high noise due to the small triangulation angle θ_1. The data of sensor 2 display low noise but are not unique, due to the small unique measurement range ΔZ_2.

However, the whole system is *more* than the sum of its parts. Since both sensors use the *same projection pattern*, they measure exactly the same surface points. This fact can be exploited! As already discussed in Sect. 5.1.1, the point clouds from both sensors can be principally compared in 3D space. But again there is a back projection-based method which is less computationally intensive - the *index-back-projection-approach* [8]. The basic work flow is shown in Fig. 5.15: The line images of the illuminated object surface (Fig. 5.15a), acquired with C_1 (Fig. 5.15b) and C_2 (Fig. 5.15c), deviate from a straight line, depending on the used triangulation angle. Due to the small θ_1, the lines in the image of C_1 are nearly straight, which enables *easy indexing* with the standard method based on 'regions of uniqueness' (see Sect. 4.1). In Fig. 5.15b the evaluated line indices in the camera image are illustrated by a color code. 3D data evaluation from C_1 delivers unique data with high noise (Fig. 5.15d).

The lines in the image of C_2 cannot be indexed uniquely (no color code). Since N_L and θ_2 are large, the regions of uniqueness display a wide overlap on the chip resulting in a high ambiguity level. The evaluation delivers highly ambiguous 3D data with a large amount of 'phantom 3D models' behind and in front of the true model. The benefits of both 3D models (low noise and unique data) are combined via 'index back projection'. For this, the 3D data from sensor 1 (Fig. 5.15d), *together with their index information*, are back projected onto the chip of C_2. The result is shown in Fig. 5.15f: If an accurate calibration as well as certain geometrical conditions (see Sect. 6.2.3) are ensured, the indexed (noisy) back projections *overlap* with the imaged line signals on the chip of C_2. Eventually, the back projected indices are assigned to the corresponding line in the image of C_2 (Fig. 5.15g). The final 3D evaluation delivers a *unique 3D model* with *precise data*.

With the introduced *index-back-projection-approach*, it becomes possible to *guarantee* correct indexing for a large number of projected lines without sacrificing precision or measurement range. For a proof of the principle, a prototype setup of the *single-shot 3D movie camera* was realized. The setup comprises the components shown in Fig. 5.14. The measurement field is $\Delta X \times \Delta Y = 300 \times 200\ \text{mm}^2$, and the effective measurement range is $\Delta Z_e = \Delta Z_1 = 100$ mm, which is limited by the depth of field of the used projector.

[9] At least for the telecentric case, which is used for the theoretical considerations. For non-telecentric setups, there is a small perspective effect, which is negligible in most cases. It can be avoided by the application of a beam splitter.

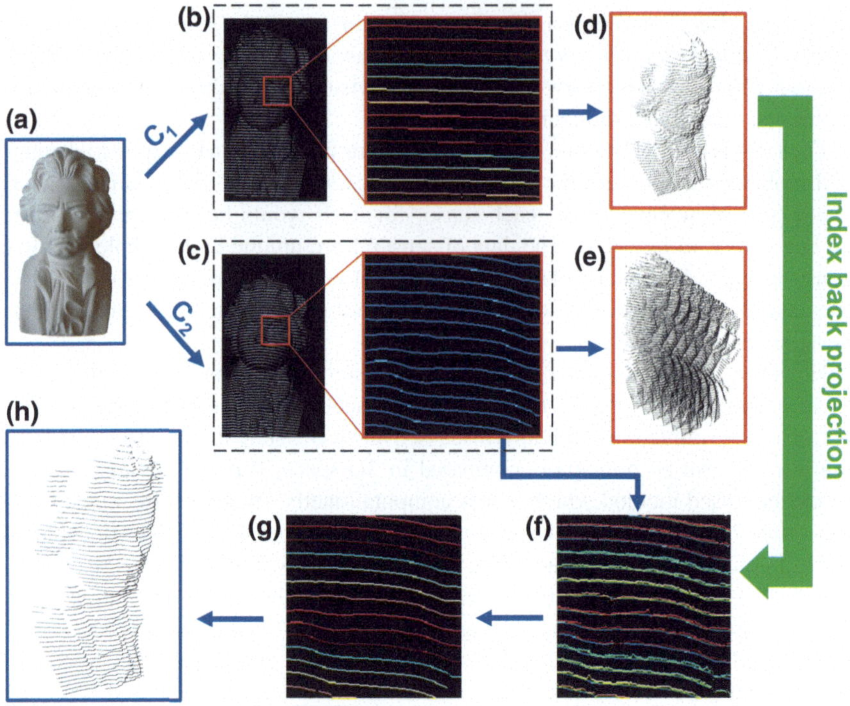

Fig. 5.15 Work flow of the index-back-projection-approach. The correctly indexed, but noisy 3D model (**d**), evaluated from C_1 (**b**) is (together with its index information) back projected onto the chip of C_2 (**f**), where the index information is assigned to the line signals (**g**). The dense 3D model (**h**), evaluated from (**g**) displays low noise and correctly indexed data

The maximum number of projected lines that could be realized with the setup is $N_L = 160$. This relates to a *ninefold ambiguity*. The theoretical considerations of the next chapter will reveal that $N_L = 160$ is already very close to the number of lines that can be *maximally* projected, if the sampling theorem is considered correctly. So *the prototype setup already works at the limit which is physically and information theoretically reasonable*.

A detailed analysis of the principle under physical and information theoretical aspects as well as more sensor specs, will be given in the next chapter.

With the prototype setup several single-shot measurements of different objects were performed with different numbers of projected lines. A selection of 3D models is shown in Fig. 5.16. Note, that all 3D models that will be shown for the rest of this thesis, consist exclusively of ***raw data*** which are *directly measured with the sensor*. No post processing, like smoothing or interpolation, is used conceal visual imperfections. Each displayed point is a directly and independently measured 3D point.

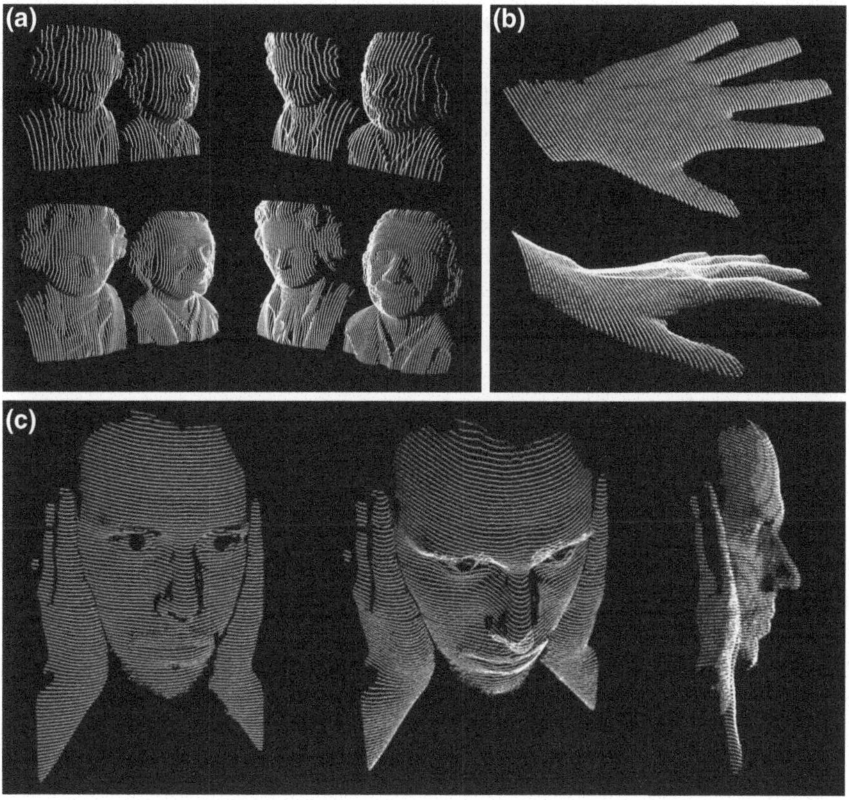

Fig. 5.16 Single-shot measurements performed with the single-shot 3D movie camera prototype setup, shown from different perspectives. **a** Plaster busts (Beethoven and Einstein), measured with $N_L = 80$ (top) and $N_L = 160$ (bottom) lines. **b** Human hand, measured with $N_L = 160$ lines. **c** Human face, measured with $N_L = 142$ lines

To record a *3D movie* of moving objects (rigid **and** non-rigid), each camera simply acquires a *temporal series of images*. Eventually each pair of simultaneously acquired images is evaluated, independent from other images in the temporal sequence. The projection remains constant during the whole acquisition process. The result is a 3D point cloud that moves in time and that can be freely observed from each viewpoint - a *free-viewpoint 3D movie*.[10] Such 3D movies will be shown in Chap. 9.

[10]Note: a *free-viewpoint 3D movie* is not related to cinema '3D movies', which should be better considered as *'2.5D movies'*. Such movies only simulate a 3D impression by offering the audience two slightly different pictures, which represent the different pictures in each eye. No real 3D data are present. The 3D evaluation is performed in the human brain. Moreover, the viewpoint of each scene is always fixed and cannot be chosen freely.

5.3 Summary and Discussion

In this chapter a new method for the solution of the profound indexing problem in multi-line triangulation was introduced. The method is based on *additional perspective information*, acquired with one or more *additional cameras* and it comes mainly in two variations: The *signal-back-projection-approach* and the *index-back-projection approach*. Although the *index-back-projection approach* is considered as an advancement of the *signal-back-projection-approach*, each variation has its own benefits.

The statistic-based *signal-back-projection-approach* is an effective method for outlier detection and correction in Flying Triangulation. Each Flying Triangulation sensor is considered as a *fixed setup* whose hardware components and geometry cannot be changed easily. The low 3D data density in each single-shot ensures enough 'space' for the back projections of outliers on the chip. This leads to good results even if only *two* cameras are applied.

It becomes obvious, that the *signal-back-projection-approach* has furthermore the potential to *improve other existing line triangulation sensors* as well. The only requirement is a second built-in camera,[11] which is already present in many commercially available sensors for the purpose of color texture acquisition. The *signal-back-projection-approach* itself does then not require additional hardware or hardware changes. The method is *purely software based*, i.e. existing sensors can be easily 'software-upgraded'.

In the second part of the chapter, the potential of back projection methods with regard to the development of a new sensor principle, the *single-shot 3D movie camera*, was discussed. Here, the situation is different. If the statistic-based *signal-back-projection-approach* would be further pursued, the high 3D data density in each single-shot would require a large amount of cameras to compensate for outliers.

A solution is introduced by the *index-back-projection-approach*, which in fact combines the advantages of two independently measuring triangulation sensors (high precision **and** large measurement range with only two cameras) without suffering their disadvantages. The combination is possible, since both sensors use the *same projection*. This means that they independently measure exactly the same surface points. The special geometrical arrangement *guarantees* correct indexing inside the whole effective measurement range. Therefore, the *index-back-projection-approach* is *deterministic*.

Both introduced back projection approaches *preserve all valuable features of multi-line triangulation*: First, they are still single-shot. Second, the back projections can be performed pointwise, which preserves the high feature resolution.[12] Moreover,

---

[11] To be precise, the triangulation angle, enclosed by the second camera and the projection should not be zero. Unfortunately this is approximately the case in the Flying Triangulation face scanner, where the signal-back-projection-approach has only a reduced effect. See [1, 6] for more in information.

[12] As long as the back projections of 'correct points' are landing inside the Gauß-correlation area of the corresponding line, no additional neighborhood information is exploited.

the principle does not require a switchable projector. A simple slide-projector, comprising a chrome on glass mask with lines is sufficient.[13] Such projectors can easily be driven by high-power light sources, which enables very short exposure times for each single-shot 3D model. Therefore, the *single-shot 3D movie camera* is especially suited for the 3D acquisition of very fast processes, e.g. deforming car bodies or deploying airbags during crash tests.

Due to the large potential for future 3D applications, the remaining part of this thesis gives special attention to the *single-shot 3D movie camera* based on the *index-back-projection-approach*. The next chapter analyzes the physical and information theoretical limits of the single-shot 3D movie camera on the basis of the introduced prototype setup.

References

1. F. Willomitzer, FlyFace – 3D-Gesichtsvermessung mittels “Flying Triangulation”. Diploma Thesis, University Erlangen-Nuremberg (2010)
2. O. Arold, Flying Triangulation' - handgeführte optische 3D Messung in Echtzeit. Dissertation, University Erlangen-Nuremberg (2013)
3. F. Willomitzer, S. Ettl, C. Faber, G. Häusler, Single-shot three-dimensional sensing with improved data density. Appl. Opt. **54**(3), 408–417 (2015)
4. F. Willomitzer, S. Ettl, C. Faber, G. Häusler, Detection and correction of line indexing ambiguities in Flying Triangulation. in *Proceedings of the 114th DGaO Conference, A12* (2013)
5. F. Willomitzer, S. Ettl, C. Faber, G. Häusler, Flying Triangulation - Towards the 3D Movie Camera. in *Fringe 2013* (Springer, Heidelberg, 2013), pp. 895–898
6. M. Kuch, Neue Indizierungsmethoden für die Multilinientriangulation. Master Thesis, University Erlangen-Nuremberg (2014)
7. G. Häusler, F. Willomitzer, P. Dienstbier, C. Faber, Tomographic Triangulation. in *Proceedings of the 114th DGaO Conference, A16 (abstract only)* (2013)
8. F. Willomitzer, G. Häusler, Single-shot 3D motion picture camera with a dense point cloud. Opt. Exp. **25**(19), 23451–23464 (2017)

[13]Nevertheless, a switchable projector is included in the prototype setup to test different numbers of projected lines. It can be replaced by a chrome on glass slide projector for a ‘final version’ of the single-shot 3D movie camera.

Chapter 6
Physical and Information Theoretical Limits of the Single-Shot 3D Movie Camera

The last chapter introduced a concept for a *single-shot 3D movie camera* based on multi-line triangulation. The feasibility has been proven by first measurements with a prototype. The demonstrated measurements are single-shot *and* display a high 3D data density.

However, it was not discussed, whether the introduced concept is *optimized*, meaning that the *3D camera* is able to *work close to the physical and information theoretical limits of 3D sensing*.

In this chapter the following aspects related to this question will be investigated:

- What is the maximal possible 3D data density? Does the sensor reach this density?
- Which measurement uncertainty can be achieved?
- Does the method preserve high object frequencies?
- Is the evaluation procedure based on the *index-back-projection-approach* optimal?
- Is the introduced combination of a *small and a large angle* optimal?
- How 'channel-efficient' is the 3D movie camera compared to other well-established methods?

As discussed, the principle of the *3D movie camera* is freely scalable, enabling the measurement of objects with sizes from several millimeters to meters (or even larger, if sufficient light is provided by the projection). For the sake of comparability, the theoretical considerations in this chapter will all be based on the hardware configuration of the *prototype setup*, which has been introduced in the last chapter. Table 6.1 gives a list of specifications, relevant for the calculation of different limits. Real measurements, performed with the prototype setup, will then reveal whether the sensor principle is really able to reach the calculated limits. A full list of specifications, also including the outcome of the following analysis, is given in the appendix (Table A1).

F. Willomitzer, *Single-Shot 3D Sensing Close to Physical Limits and Information Limits*, Springer Theses,
https://doi.org/10.1007/978-3-030-10904-2_6

Table 6.1 Hardware specs of the prototype setup

Measurement field[a,b]	$\Delta X \times \Delta Y$	300×200 mm^2
Effective measurement range	$\Delta Z_e \leq \Delta Z_1$	100 mm
Stand-off distance[a,c]	w	∼ 500 mm
Triangulation angle[a] sensor 1	θ_1	∼ 1°
Triangulation angle[a] sensor 2	θ_2	∼ 9°
Number of projected lines	N_L	≤ 160
Number of active pixels[b]	$N_x \times N_y$	1024×682
Pixel size	$(d'_{pix})^2$	(4.65 μm)^2
Numerical observation aperture[a]	$\sin u_{obs}$	0.00225
Numerical illumination aperture[a]	$\sin u_{ill}$	0.0075
Mean illumination wavelength	λ	550 nm

[a]in the middle of the measurement range
[b]total overlap of projection and observation
[c]measured from the first objective lens

6.1 Data Density

In this section, the maximal reachable data density for practical multi-line triangulation setups is discussed. Eventually, the result will be compared with the data density of the 3D movie camera. For the following discussions, it is assumed that according to the definition of Chap. 2, each pixel displays widely independent data.

For multi-line triangulation a *maximal data density* means a *maximal number of projected lines* N_L. The fact that a binary line signal cannot be projected arbitrarily dense has already been discussed. The first restriction of N_L is given by the sampling theorem. Considering the *Nyquist-frequency* in Eq. (2.22), one could naively think that a line number of $N_L \approx N_x/2$ is possible, which would result in roughly 500 lines for the given sensor specifications.

However, such a number cannot be reached in practice. First, the *Nyquist-limit* is strictly speaking only valid for infinitely large object and chip sizes. Second, the precise detection of the line intensity maximum needs additional space. The state of the art method relies on the approximation of the signal shape $I(x')$ (row-wise cross section through each line signal) with a Gauß-function[1]

$$I(x') = I_0 \cdot \exp\left(-\frac{1}{2}\left(\frac{x' - x'_0}{\sigma}\right)^2\right) . \tag{6.1}$$

In addition to the desired sub-pixel precise position of the intensity maximum x'_0, Eq. (6.1) has two more unknowns: The maximal intensity I_0 and the standard

[1]One could ask why a *Gauß* function is used for approximation instead of e.g. a *sinc* function. The answer is that the *Gauß* function approximates the signal shape best in practice, because most objectives are not diffraction limited.

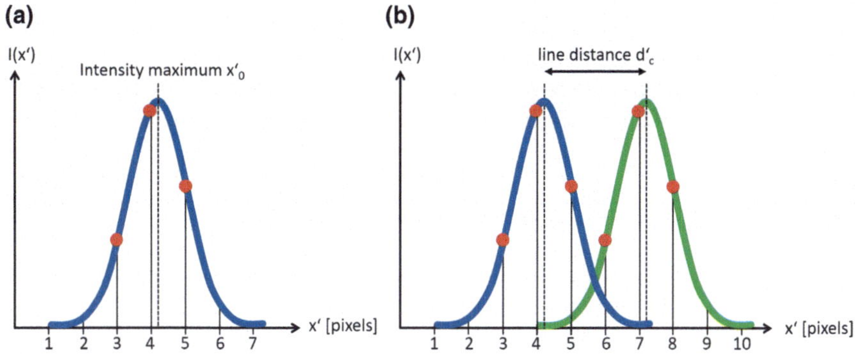

Fig. 6.1 Sub-pixel precise evaluation of line profiles via Gauß-approximation. **a** Localization of the line intensity maximum. **b** Minimal distance of two neighbored lines

deviation σ. This means the gray values of at least *three* pixels have to be evaluated [1].

To enable a robust sub-pixel interpolation, according to [1, 2], one of the chosen pixels should display the maximal *measured* intensity, whereas the other two pixels (left and right) should display intensities close to one half of the maximum intensity. Figure 6.1a schematically illustrates this case. It is to be seen that the resulting minimal linewidth on the chip should be approximately (and not much larger than) 4 pix. According to Fig. 6.1b, two sub-pixel interpolated lines remain distinguishable, if the distance of their maxima on the camera chip is *at least* $d_c' \geq 3\ pix$. This would result in $N_L = N_x/3 \approx 340$ lines for the prototype setup.

However, this limit could only be reached for flat surfaces. If a possible surface inclination as well as the triangulation angle between projection and camera is considered, N_L is further decreased due to perspective contractions. Figure 6.2 illustrates this situation: Two lines with distance d_p are projected onto a surface which is tilted by the angle α and observed with a camera under the triangulation angle θ. Under the

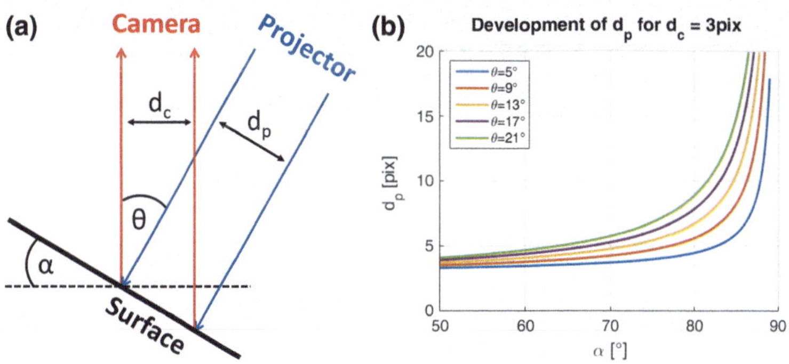

Fig. 6.2 Perspective contraction of line distances at tilted surfaces

shown angular conditions, the line distance d_c, seen by the camera, becomes *smaller* due to the perspective contraction. For a sensor setup able to measure steep object parts as well, the projected linewidth d_p has to be chosen *sufficiently large*, so that $d'_c \geq 3\ pix$ is always met. According to Fig. 6.2, d'_c can be calculated as

$$d'_c = d'_p \cdot \frac{\cos\alpha}{\cos(\alpha - \theta)} \overset{!}{\geq} 3\ pix\ . \tag{6.2}$$

Assuming a maximal allowed inclination of $\alpha \approx$ 80–85° and a triangulation angle of $\theta = \theta_2 \approx 9°$, d'_p has to be chosen in a range of $d'_p \approx 6–8\ pix$.

The introduced prototype setup is tailored to work at this limit. The setup projects up to $N_L = 160$ lines. Note that the projector differs slightly in terms of field, resolution and aspect ratio from the observation. Moreover, the optical axes of all components are slightly tilted to each other. This results in the fact that the fields of projector and cameras do not overlap perfectly in the whole measurement volume. Effectively, $N_L \approx$ 140–150 lines are visible within the field of the cameras, which relates to a distance of projected lines of $d'_p \approx 7\ pix$. This is illustrated in Fig. 6.3: Fig. 6.3a shows the original projector-pattern (with a magnification window) of 160 straight narrow lines. Figure 6.3b presents the image of an planar untilted plate,[2] illuminated with the pattern of Fig. 6.3a and imaged with C_1. Since $\theta_1 \approx 1°$ and $\alpha = 0°$, it follows from Eq. (6.2) that $d'_c \approx d'_p$ in this case. So d'_p can be directly determined from Fig. 6.3b. As shown in the magnification window, the line distance is $d'_p \approx 7\ pix$. The red/green colored bar visualizes the pixel size for a comparison.

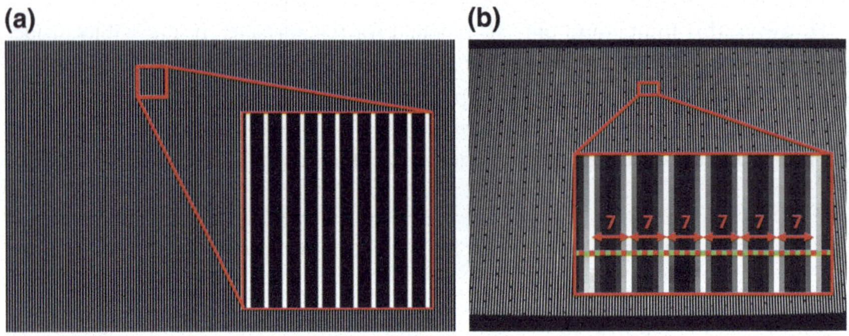

Fig. 6.3 Evaluation of the effective projected line distance. **a** Original projector-pattern of 160 straight narrow lines projected onto the object. The magnification window (red) illustrates, that only equidistant, straight, narrow lines are projected. **b** Image of an untilted plate with on-projected pattern taken with C_1. The effective projected line distance can be directly read off (see red magnification window)

[2]The image of Fig. 6.3b is originally used for the calibration of the sensor. With the black dots on the plate, its position in space is determined. More information about the calibration to be found in Sect. 8.1.

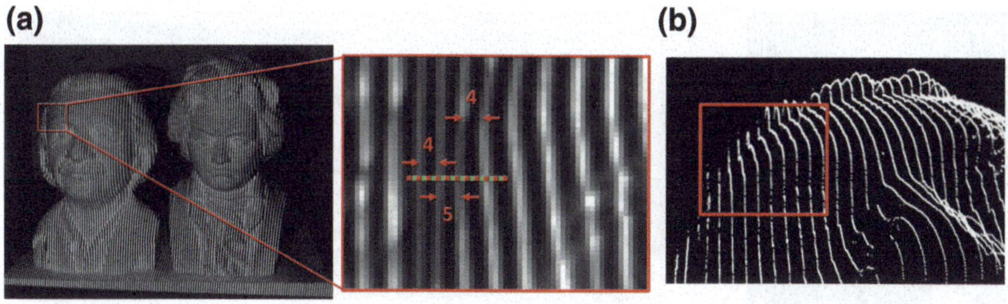

Fig. 6.4 Perspective contraction of the line distance in real measurements for $N_L = 160$ projected lines. **a** Camera image to be evaluated, containing perspective contractions at steep surface parts. **b** Evaluated 3D model

Figure 6.4 demonstrates the contraction by perspective during real measurements: For the camera image to be evaluated (camera image of C_2 in Fig. 6.4a), the line distance is reduced down to $d_c \approx 4\ pix$ for steep surface parts, even though a line distance of $d'_p \approx 7\ pix$ is projected. The related part of the evaluated 3D model is shown in Fig. 6.4b: The lines can still be evaluated separately although the data display higher noise according to the reduced brightness of the lines. It becomes clear that the absolute limit of the line density is close to being reached or is reached already. Indeed, measurements with a projected line number of $N_L > 160$ are only possible at relatively untilted surface parts.

Summing up, the maximal number of $N_L = 160$ lines, projected by the prototype setup is already *close to the theoretical limit* under the following conditions:

- Sub-pixel precise evaluation of the line intensity maximum by Gauß approximation
- Objects with steep surface parts.

The two conditions are necessary for a precise measurement of a broad variety of surface types. The given line number is in agreement with experimental results. For $N_L > 160$ projected lines, tilted surface areas become unmeasurable. Hence, it can be said, that:

> The prototype setup of the *single-shot 3D movie camera* acquires 3D data with a density *close to the theoretical maximum*!

Figure 6.5 displays a single-shot measurement of a human face with $N_L = 160$ projected lines. The high data resolution in perpendicular line direction of around $(\delta x_{dat})^{-1} \approx (2\ \text{mm})^{-1}$ is seen in the image of C_2 (Fig. 6.5a) as well as in the raw data of the evaluated 3D model (Fig. 6.5b).

Note that for 'everyday applications', which do not require ultra dense 3D models, it might be sufficient to project a slightly lower number of around $N_L \approx 140$ lines. This makes the principle less prone to possible decalibration and a broadening of the line signal caused by motion or volume scattering (see [1, 3, 4] for more information).

In the next section, the measurement uncertainty of the *single-shot 3D movie camera* is analyzed.

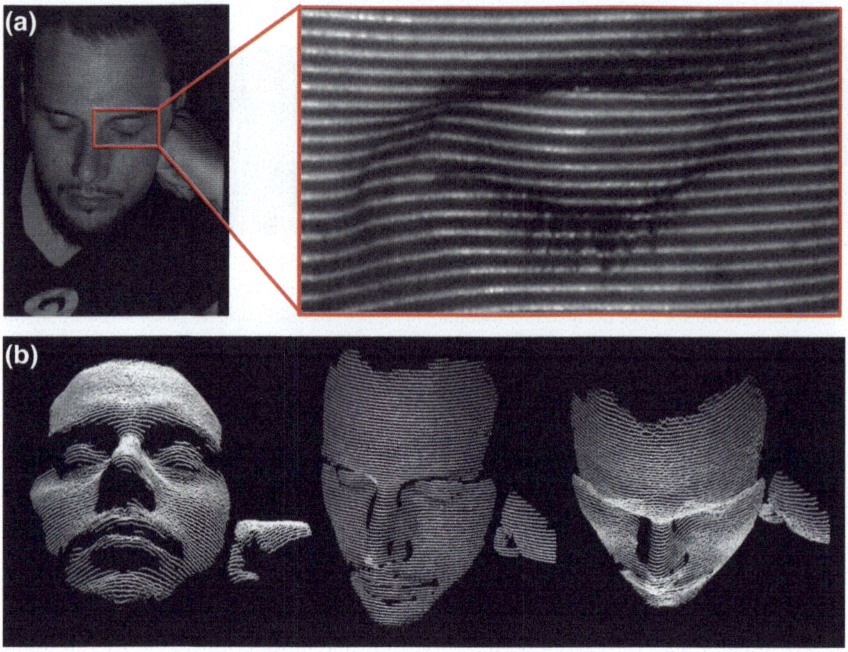

Fig. 6.5 Measurement of a human face with 160 projected lines. **a** Camera image to be evaluated (C_2). **b** Evaluated 3D model (raw data), shown from three different points of view

6.2 Measurement Uncertainty

This section discusses whether the single-shot 3D movie camera is ***physically optimized*** due to the definition in Sect. 2.2.5. This is done by estimating an upper bound for the achievable measurement uncertainty, based on the specs of Table 6.1. A comparison with measurements on different surfaces reveals, if the single-shot 3D movie camera acquires data from the object surface with a quality ***as good as physics allows***.

6.2.1 Theoretical Limit of the Measurement Uncertainty

Speckle noise puts an ultimate limit on the distance precision of line triangulation setups. This was discussed in Sect. 2.2.4 and a 'recipe' for the *physical optimization* (i.e. for the reduction of the speckle noise) was given in Sect. 2.2.5.

'Reduction of speckle noise' stands for a reduction of the speckle contrast C_s. For the fully coherent case ($C_s = 1$), the minimal achievable measurement uncertainty of the prototype setup in the center of the measurement volume is calculated with Eq. (2.12) to:

$$\delta z_{speckle}(C_s = 1) = \frac{1}{2\pi} \cdot \frac{550\ \text{nm}}{0.00225 \cdot \sin(9°)} = 249\ \mu\text{m} \tag{6.3}$$

For the calculations the wavelength λ is the approximate mean wavelength of the (white) illumination spectrum, $\lambda = \bar{\lambda} = 550\,\text{nm}$.

For an *'incoherently designed setup'*, according the 'recipe' in Sect. 2.2.5, $\delta z_{speckle}$ can be significantly reduced. This 'recipe' was taken into account for the design of the prototype setup. In the following, the known contributions to the reduction of the speckle contrast of the prototype setup are discussed separately.

Reduction of $\mathbf{c_{spat}}$:
By exploiting an illumination aperture $\sin u_{ill}$, larger than the observation aperture $\sin u_{obs}$, the spatial contribution to C_s can be reduced in accordance with Eq. (2.18) by a factor of

$$c_{spat} = \frac{0.00225}{0.0075} = 0.3\ . \tag{6.4}$$

Reduction of $\mathbf{c_{pol}}$:
In the prototype setup a *LED-DMD projector* (no LCD!) [5] is used, which provides *unpolarized light*. Moreover, most object surfaces to be measured are depolarizing. In accordance with Eq. (2.14) C_s is reduced by a factor of

$$c_{pol} = \frac{1}{2}\ . \tag{6.5}$$

Reduction of $\mathbf{c_{pix}}$:
A pixel size larger than the size of a subjective speckle would reduce the speckle contrast due to averaging effects. However, as discussed in Sect. 2.2.4 this would also reduce the lateral resolution of the setup. This is the reason why the prototype setup uses pixels with approximately the same size as the subjective speckles. Due to Eq. (2.15), the result is only a small reduction of C_s by a factor of

$$c_{pix} = \frac{550\ \text{nm}}{4.65\ \mu\text{m} \cdot 0.125} = 0.95\ . \tag{6.6}$$

Here, the image sided numerical aperture is calculated with the f-number ($\#_f \approx 4$) of the camera objective to $\sin u'_{obs} \approx 1/2\#_f = 1/8$.

Reduction of $\mathbf{c_{temp}}$:
As shown in Eq. (2.16), the contribution of the temporal coherence to C_s is dependent on the relation of the surface height variation δh within one diffraction disc and the coherence length l_c of the illumination. This relation is unknown and could be only determined experimentally. In the next section, the measurement uncertainty is systematically evaluated on a planar screen, laminated with white paper (calibration

plate). Since this screen displays volume scattering, an experimental determination of δh requires intensive effort and would only deliver a rough estimation.

Therefore, the reduction of c_{temp} is excluded from the theoretical considerations for the moment. The influence of the temporal coherence will be discussed after the presentation of the experimental results in the next section.

Based on the theoretical reductions of c_{spat}, c_{pol} and c_{pix}, it is nevertheless possible to formulate an *upper limit* for the reduction of the speckle contrast:

$$C_s < c_{spat} \cdot c_{pol} \cdot c_{pix} = 0.3 \cdot 0.5 \cdot 0.95 \approx 0.15 \tag{6.7}$$

The related minimal achievable measurement uncertainty in the middle of the measurement range of the prototype sensor is:

$$\delta z_{speckle}(C_s = 0.15) = \frac{0.15}{2\pi} \cdot \frac{550\ \text{nm}}{0.00225 \cdot \sin(9°)} = 37.3\ \mu\text{m} \tag{6.8}$$

This means, that even without the contribution of c_{temp}, the precision of the sensor can be improved to a value of only 15% of the fully coherent case. Naturally, as a consequence of such a large reduction, the importance of other sources of noise—such as electronic noise or shot noise—may become important as well. In that case, the sensor's precision is no longer dominated by the speckle noise.

So much on the theory; the next section offers concrete experiments proving the estimated noise reduction actually can be achieved.

6.2.2 Experimental Evaluation of the Measurement Uncertainty

In this section the precision of the single-shot 3D movie camera prototype setup is determined for two different surface types. First, the precision is evaluated for measurements on the laminated calibration plate, which is positioned at different depths in the measurement volume. Since the measurements are reproducible and many samples can be taken, this precision evaluation is quite systematic and reliable for the respective surface type. In a second evaluation, the precision of the sensor 'in action' is determined. For this, different line samples from single-shot measurements on human faces are evaluated.

Measurement Uncertainty on the Calibration Plate

The calibration plate of the sensor is made of so-called 'Aluminum-Dibond' and is coated with an adhesive paper with an imprinted dot pattern. During the calibration the dot pattern is used to determine the position of the plate in space. More information about the plate and the sensor calibration is given in Sect. 8.1 and [6].

The process of the precision evaluation is visualized in Fig. 6.6: A pattern with $N_L = 142$ lines is projected onto the calibration plate (see Fig. 6.6a), which is placed approximately perpendicular to the optical axis of the projector. An image of the illuminated plate is acquired at 11 different positions within the measurement depth (see Fig. 6.6b). The plate is placed by hand, the exact positions are unknown. Note that the ***distance precision*** δz of the sensor *should not be affected by the calibration*. This is the reason why 3D data, from *direct 3D evaluation* via calibration transformation, are ***not used*** for the precision evaluation on the calibration plate. Only the 'physical' noise of the lines is considered.

For the noise evaluation, the sub pixel precise intensity maxima are determined row-wise for each line in each of the 11 images. Eventually, the tilt of each line is

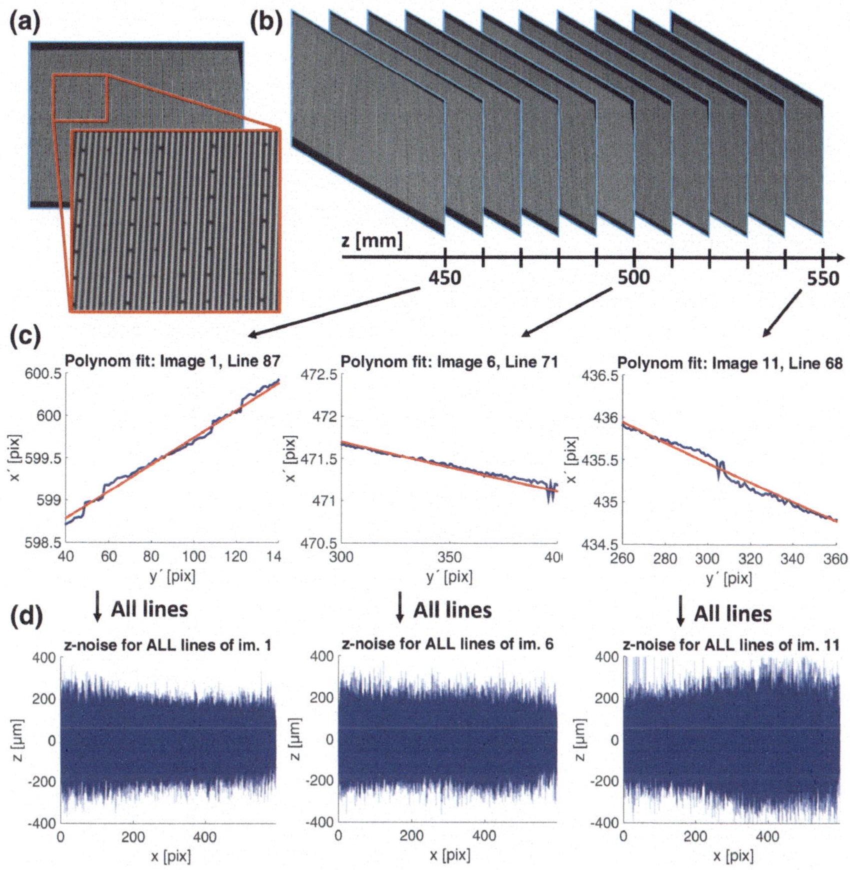

Fig. 6.6 Evaluation of the depth precision of the prototype setup on the calibration plate. **a** Representative image from one depth with magnification window. **b** Image stack of 11 images acquired at 11 depths within the measurement range. **c** Representative segments of single lines in chip coordinates. Blue: Positions of the evaluated intensity maxima. Red: Polynomial fit of 5th order. **d** Superimposed z-noise (peak-to-valley) of all lines from one image

compensated by subtracting a polynomial fit of 5th order. Line segments that intersect a dot on the calibration plate are not further considered for the evaluation. Figure 6.6c shows exemplarily three sections (each with 100 pix length) of single lines, taken from the beginning (450 mm), the middle (500 mm) and the end (550 mm) of the measurement range. The red curve displays the polynomial fit, while the blue curve displays the evaluated maxima positions—still in lateral chip-coordinates.

After subtraction of the polynomial fit curve, only the lateral variation of the line intensity maximum (the noise in x-direction) remains. From this noise, the standard deviation $\delta x'$ is calculated—again for *each line* in *each image*.

Eventually, $\delta x'$ is transferred into a depth value via the triangulation formula (Eq. (2.1)):

$$\delta z = \frac{\delta x'}{\beta' \sin\theta} \tag{6.9}$$

Equivalent to the procedure described above, δz could also be determined by the standard deviation of the peak-to-valley 'z-noise', which can be calculated by projecting the lateral variation of the line intensity maximum in 3D space using the triangulation formula. Figure 6.6d shows the superimposed z-noise of all lines from one image in one plot. As expected, it is to be seen that the noise is statistically distributed and forms a certain 'noise pipe'. The mean standard deviation can already be roughly estimated from the plot.

The result of the evaluation is shown in Fig. 6.7 and Table 6.2. Note that the triangulation angle and the object sided field width (which affects β') differs across

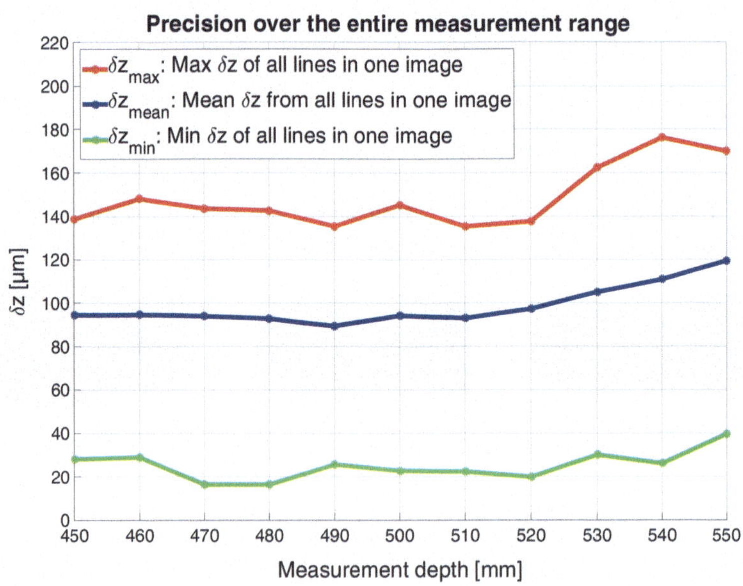

Fig. 6.7 Depth precision of the prototype setup on the calibration plate

Table 6.2 Depth precision of the prototype setup on the calibration plate. Determined values and related triangulation angles and field widths for three different measurement depths

	z = 450 mm	…	z = 500 mm	…	z = 550 mm
θ	10.1°	…	9.1°	…	8.2°
ΔX	270 mm	…	300 mm	…	330 mm
δz_{max}	139 μm	…	146 μm	…	170 μm
δz_{mean}	95 μm	…	95 μm	…	120 μm
δz_{min}	29 μm	…	23 μm	…	40 μm

the measurement range. This is taken into account for the calculations. Also note, that *whole* lines over the whole (vertical) field are considered, not only specific line segments.

For each depth (i.e. acquired image), the minimal value δz_{min} ('best line'), the maximal value δz_{max} ('worst line') and the mean value δz_{mean} of all lines in the respective image is given. The evaluated measurement uncertainty is better than $\delta z_{max} \leq 180\,\mu\text{m}$ in the whole measurement volume. Moreover, the mean value of all lines is better than $\delta z_{mean} \leq 120\,\mu\text{m}$. If only the 'best' line in each image is considered, the related measurement uncertainty is even below $\delta z_{min} \leq 40\,\mu\text{m}$ within the whole measurement volume.

How do these results fit into the theoretical considerations of the prior section?

Besides the physical restrictions of (speckle) noise, the uncertainty in distance localization is also restricted by many 'technical' factors. One very important technical factor is the determination of the line intensity maximum via Gauß approximation. Due to [7], the precision of this process decreases with the width of the line signal. This is the reason why the best values δz_{min} always originate from lines in the middle of each image, where the lines are narrow, since the influence of aberrations is small.[3]

If, as in the previous section, *only the fundamental physical restrictions are of interest*, it is reasonable (as an exception) to consider the *best* values of the measurement uncertainty. In the middle of the measurement range, the best value $\delta z_{min} = 23\,\mu\text{m}$ is ***over ten times smaller*** than the theoretically calculated value of $\delta z_{speckle}$ for the fully coherent case ($C_s = 1$). The absolute best value $\delta z_{min} = 17\,\mu\text{m}$ is measured at a depth of 470 and 480 mm. This is the result of the physical optimization that leads to a reduction of the speckle contrast.

The theoretical calculation of the speckle contrast reduction in the prior section delivered an upper bound for the best value in the middle of the measurement range (500 mm) of $\delta z_{speckle} = 37.3\,\mu\text{m}$. The experimental evaluation in this section reveals this upper bound can really be reached and even surpassed by the prototype sensor.

This is due to c_{temp} which was not considered for the theoretical calculation so far: If speckle noise would be the *only* noise source, one could finally estimate c_{temp}

[3]The 'best' lines might be located in the middle of the horizontal field ΔX, but are still spread over the whole vertical field width ΔY, which means that they are already slightly influenced by aberrations.

to $23/37.3\ \mu\text{m} \approx 0.6$. According to Eq. (2.16), this would mean that the penetration depth δh is approximately the coherence length l_c. As the paper-coating of the plate displays volume scattering, the estimate $\delta h \approx l_c$ seems too low.

However, after the speckle noise had been significantly reduced, other noise sources play an important role as well. All of these noise sources (electronic noise, shot noise, uncertainty of Gauß-approximation, ...) contribute to the measurement uncertainty. This means, that c_{temp} is in fact much smaller than 0.6.

The other noise sources are not as easily accessible as the speckle noise. Either they are already optimized (like the shot noise) or they can just be reduced by extensive technical effort (e.g. chip cooling). Hence, it can be said that:

The single-shot 3D movie camera prototype is able to reach a ***distance precision close to the limit of what is physically possible*** *with the used (standard) hardware!*

Measurement Uncertainty on Human Skin

Although the sensor concept of the *single-shot 3D movie camera* allows for the measurement of a broad variety of objects and surface types, the introduced prototype is tailored for a specific application: The measurement of *human faces*. A human face fits perfectly into the measurement volume. Moreover, the benefits of a dense and precise single-shot 3D movie acquisition can be demonstrated without relying on elaborate hardware: A talking human face is a *non rigid moving object*, which moves *fast*, but not as fast that cameras with high frame rates or high power illumination would be required.

Hence, the measurement uncertainty on human skin is of special interest for the current sensor specification and is a good parameter to estimate the precision of the sensor 'in practice'.

A systematic precision evaluation of a series of images, as performed for the calibration plate, would require elaborate programming. The multitude of different facial expressions leads to a multitude of different looking camera images. Therefore, the measurement uncertainty on human skin is only estimated by a small sample of two single-shot images, taken from two different 3D movies (see Fig. 6.8). In a first attempt, the precision is evaluated with the same method as described above: The localization uncertainty $\delta x'$ is determined along each line and transferred into 3D space via Eq. (6.9). Since the exact position of the face in the measurement volume is unknown, the triangulation angle $\theta = 9°$ and the field width $\Delta X = 300$ mm from the middle of the measurement range are considered for the calculation.

The polynomial fit of the 3D profile, which is used as a dummy for the true surface, should follow the smooth variations of the surface, but *not* the variation of the noise. This is the reason why only lines displaying relatively low spatial frequencies can be evaluated.

For each image to be evaluated (image taken from C_2 with $N_L = 142$ projected lines) a sample set of 20 (line 7–line 26), respectively 17 (line 6–line 22) line segments

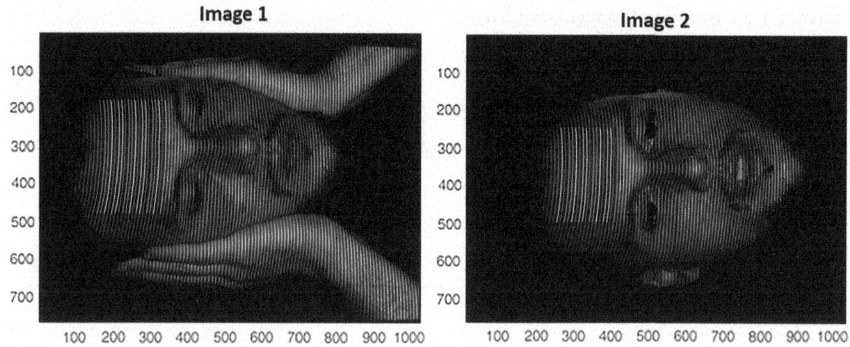

Fig. 6.8 Images for the evaluation of the depth precision on human skin: Two single-shot images, taken from two different 3D movies, are considered. The precision is evaluated along 20 (image 1) and 17 (image 2) line segments of $\sim$ 300 pix length located at the forehead of the measured person

of $\sim$ 300 pix length, located at the forehead of the measured person is considered. The two camera images with the used line segments (colored) are shown in Fig. 6.8. Exemplary plots of polynomial fits and noise from image 1 are shown in Fig. 6.9. An evaluation of the precision delivers a measurement uncertainty better than $\delta z_{max} =$ 194 µm for all considered line segments in image 1, and better than $\delta z_{max} = 203$ µm for all considered line segments in image 2. A summary of the results is given in Table 6.3.

Although the above discussed method using the triangulation formula (= 'method (a)') is the proper method for evaluating the precision, it does not reflect the procedure how 3D data are really evaluated with the sensor software. In practice, more complex calibration transformations are used instead of the simple triangulation formula. To exclude any influence of the evaluation method in use, the precision is again evaluated with a second method (= 'method (b)'): The direct calculation of the standard deviation on the 'z-noise' of the already evaluated data. A comparison of both methods is visualized in Fig. 6.9 for the same line segment (note the different axis labeling!). The result is as expected: The precision evaluation of the same sample of lines with method (b) yields approximately the same results: $\delta z_{max} = 176$ µm for all considered lines of image 1 and $\delta z_{max} = 202$ µm for all considered lines of image 2. The slight differences for image 1 might originate from a falsely estimated triangulation angle and field width for the evaluation with method (a). The full result of the evaluation is given in Table 6.3.

Summing up, the sensor displays a measurement uncertainty on human skin, which is estimated to be better than $\delta z \approx 200$ µm in the whole measurement volume. According to Eq. (6.9), this relates to an object-sided lateral localization uncertainty of $\delta x = 31.3$ µm, which corresponds roughly to $\delta x' \approx 0.1$ pix.

A comparison with the evaluated values from the calibration plate in Table 6.2 reveals, that δz_{max} is only slightly worse on human skin. For δz_{mean} and δz_{min} the

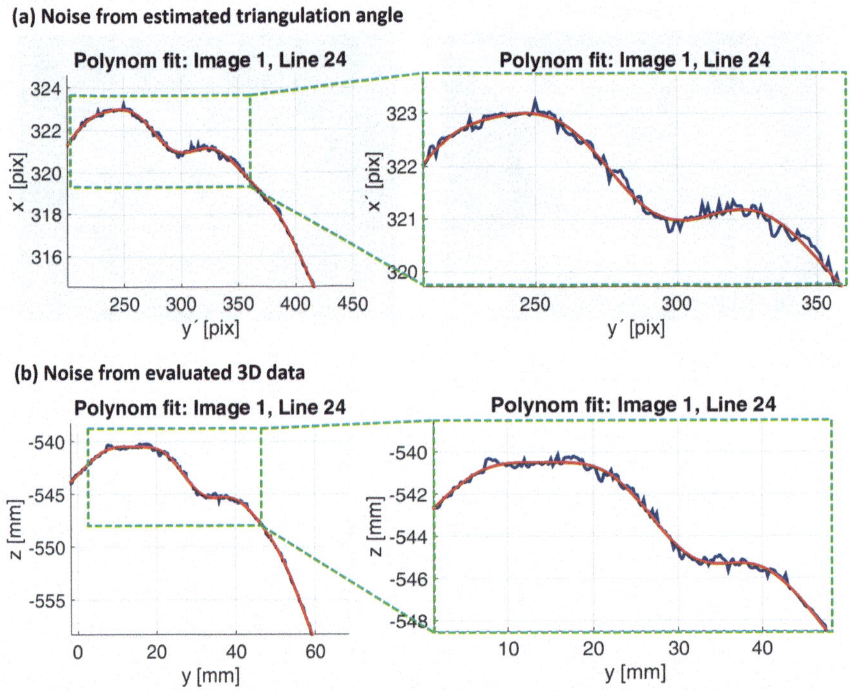

Fig. 6.9 Evaluation of the depth precision on human skin with two different methods. Method (**a**): 'Noise from estimated triangulation angle' consisting of the evaluation of $\delta x'$ on the chip and calculation of δz via Eq. (6.9). Method (**b**): 'Noise from evaluated 3D data' consisting of the evaluation of δz from the z-noise of 3D data, delivered by the direct calculation with the 3D evaluation algorithm

Table 6.3 Depth precision on human skin evaluated with two different methods

		Image 1	Image 2
Method (a)	δz_{max}	194 μm	203 μm
	δz_{mean}	170 μm	165 μm
	δz_{min}	135 μm	127 μm
Method (b)	δz_{max}	176 μm	202 μm
	δz_{mean}	151 μm	171 μm
	δz_{min}	118 μm	136 μm

evaluated values on human skin are much worse. This might be surprising at the first glance: For visible light human skin is an *intensive volume scatterer*. Due to Eq. (2.16), this should largely reduce the temporal speckle contrast, which should lead to a better precision. However, strong volume scattering also leads to a *large increase of the line width* in the camera image. Due to the reduced precision of the intensity maximum evaluation of each line [7], the good values for δz_{mean} and δz_{min} from the calibration plate cannot be reached on human skin.

6.2.3 *Influence of the Triangulation Angle*

As discussed in Chap. 2, the triangulation angle θ plays an important role for the depth precision of each (line) triangulation sensor: It converts the lateral localization uncertainty of the light signal $\delta x'$ on the chip into a longitudinal distance uncertainty δz in space via Eq. (6.9). For a good distance precision, the triangulation angle should therefore be chosen as large as possible. This section deals with the theoretical and practical restrictions of the triangulation angle.

The *single-shot 3D movie camera* prototype employs two cameras, which enclose two triangulation angles with the projection. The precision of the output data is defined by the (larger) triangulation angle θ_2 of the second camera. Due to the special geometry of the system, θ_2 cannot be chosen arbitrarily. Noise has to be taken into account. This can be explained as follows: After the first 3D data evaluation from C_1, the precision of the *unique but noisy* 3D model is given by

$$\delta z_1 = \frac{\delta x_1}{\sin \theta_1} \, . \tag{6.10}$$

This model is now back projected onto the chip of C_2. Due to the noise on the data, the back projections do not land exactly on the line maxima in the image, even if a *perfect* calibration is assumed. The 'scattering' $\Delta S'$ of the back projections around the line maxima is given by

$$\frac{\Delta S'}{2} = \delta z_1' \cdot \sin \theta_2 = \delta x' \, \frac{\sin \theta_2}{\sin \theta_1} \, , \tag{6.11}$$

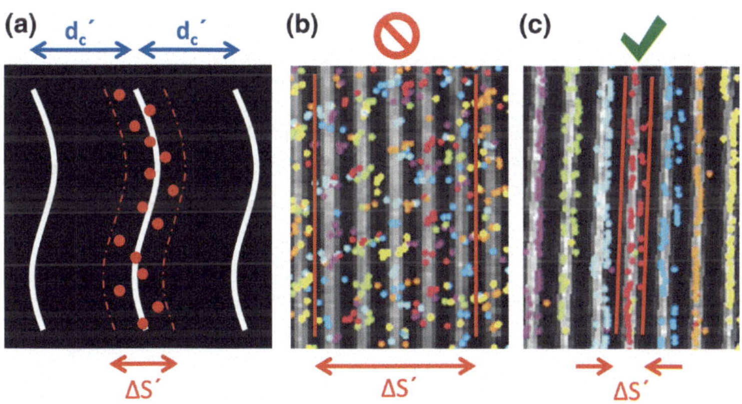

Fig. 6.10 Scattering $\Delta S'$ of back projected 3D points in the camera image of C_2. For a unique assignment of a back projected index to a line, $\Delta S'$ has to be smaller than the line distance d_c' on the chip. **a** Schematic visualization: White lines and red dots as back projections. **b** Large scattering. No unique index assignment. **c** Unique index assignment due to small scattering

where $\delta x_1' = \delta x_2' = \delta x'$ being considered to be equal for both cameras. As visualized in Fig. 6.10, a *unique assignment* of the back projected index to a line is only possible, if $\Delta S'$ is *smaller* than the distance d_c' between two lines in the camera image (or more precise: $\pm\frac{d_c'}{2}$):

$$d_c' \overset{!}{>} 2\,\delta x' \,\frac{\sin\theta_2}{\sin\theta_1}\,. \tag{6.12}$$

Figure 6.10b, c visualize two sections of images from C_2. The back projected 3D points are plotted in the respective camera image, the different indices are visualized by different colors. By comparing the scattering of the back projections with the 'red' index in both image sections, it is obvious that Fig. 6.10b does not meet the condition for unique index assignment.

What does this mean for the prototype setup?

For the correct indexing of maximal $N_L = 160$ lines within a measurement depth of $\Delta Z_1 = 100$ mm and a field width of $\Delta X = 300$ mm, Eq. (4.4) delivers that θ_1 should be not larger than $\theta_1 \approx 1°$. According to Eq. (6.12) the 'scattering' $\Delta S'$ of back projections always has to be smaller than the line distance d_c' in the camera image. As shown in Sect. 6.1, the line distance in the camera image can decrease due to perspective contraction. In order to ensure the separability of lines, the minimal allowed line distance on the chip is $d_{c,min}' = 3\ pix$.

This means, *the triangulation angle θ_2 has to satisfy **two** conditions for all possible surface inclinations α*: First, $d_c' > \Delta S'$, according to Eq. (6.12), and second $d_c' > 3\ pix$, according to Eq. (6.2). These conditions put an upper limit on θ_2, which is dependent on α. Figure 6.11 presents a graphical visualization of both conditions in one plot. For the calculation, the fixed parameters of the prototype setup in Table 6.1 and the 'worst' measured lateral localization uncertainty $\delta x = 31.3$ μm are used, while θ_2 is varied as a function α.

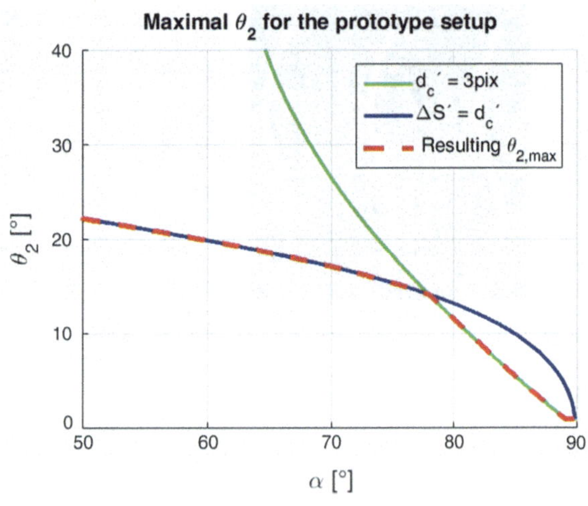

Fig. 6.11 Maximal possible triangulation angle θ_2 for the prototype setup

The blue graph represents the maximal possible θ_2 that meets Eq. (6.12), whereas the green graph represents the maximal θ_2 for Eq. (6.2). It can be seen that for inclinations up to $\alpha \approx 78°$ the 'scattering-condition' of Eq. (6.12) puts an upper limit on θ_2. The remainder of the blue graph represents values for θ_2 with the 'scattering-condition' still being satisfied, but the line distance in the image of C_2 already being smaller than 3 pix. Hence, the second condition of Eq. (6.2) puts an upper limit on θ_2 for all higher surface inclinations. For $\alpha \approx 83°$, the calculation delivers $\theta_2 \approx 9°$, which is the angle used in the prototype setup. This is not surprising, since the number of projected lines N_L was explicitly chosen under this condition. In case a larger θ_2 is desired (e.g. for higher precision), the consequence is a *decrease of* N_L if high inclinations should still be measurable.

If the measurement of high surface inclinations is not important, θ_2 could be chosen larger—in theory up to $\sim 20°$ for the prototype setup (see Fig. 6.11).

However, such large values are impracticable. There are technical restrictions such as *device compactness*, or -of much more significance- the *occlusion of surface areas*. Especially surface occlusions caused by a large triangulation angle can drastically decrease the number of measured points. Another technical restriction is the *accuracy of the calibration*: Due to the large lever of sensor 1 (θ_1 is only $\sim 1°$), a slight decalibration can result in a significant shift of the back projections on the chip of C_2 of *several pixels*. The prototype setup is especially prone to decalibration, since it comprises an off-the-shelf projector [5], which suffers from unstable mechanics and thermal drift.

6.3 Feature Resolution

The capability of line triangulation systems to resolve small object features has already been discussed several times. Due to the considerations in Sect. 2.2.2, line triangulation is principally able to provide the best possible feature resolution ('pixel-dense') in line direction. Perpendicular to the line direction, a small correlation area of at least ~ 3 pix is required for sub pixel precise line maximum detection.

However, all these considerations are limited by the sampling theorem. All sensors (even PMT or line triangulation) display artifacts at sharp edges, where the sampling theorem is violated. The main problem for line triangulation is the deformation of the line intensity profile at a sharp edge. Related discussions can be found, e.g. in [1, 3, 8]. If the class of objects to be measured should not be restricted, such small artifacts seem to be unavoidable. In [1], a method for the detection of deformed intensity profiles is proposed, that enables for the compensation of artifacts—however, not without exploiting additional neighborhood information.

Nevertheless, line triangulation is able to preserve high object frequencies much better than other single-shot principles due to the absence of spatial encoding strategies. An example is shown in Fig. 6.12: The object is a folded piece of paper. The lines are projected approximately perpendicular to the edges, which reduces the mentioned artifacts [3]. A 3D video was captured while the paper folds were con-

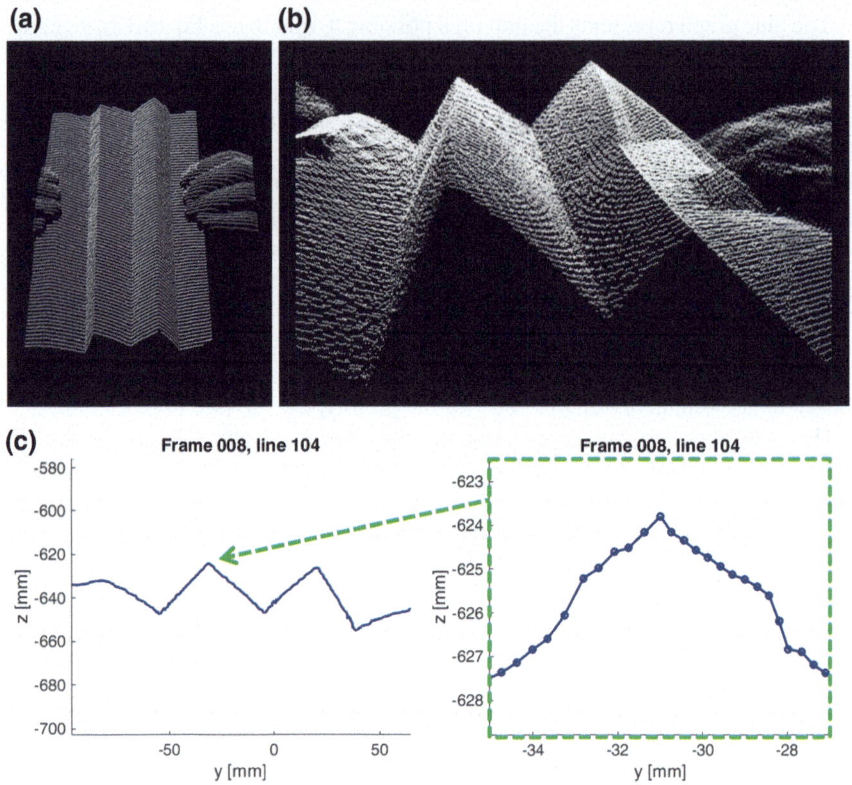

Fig. 6.12 3D measurement of a folded piece of paper (raw data). **a** and **b** Single frame of the acquired 3D movie from two different perspectives. **c** 3D data of a single line profile magnified within the field of $\sim 6\times\ 6\ \text{mm}^2$

tinuously stretched and compressed (like an accordion). The full video can be found on [9, 10] (direct link: tinyurl.com/3DMovCam07).[4] Figure 6.12a, b display a single frame of the video (raw data), shown from two different viewpoints. In compliance with the discussions in Chap. 2, the 'roof edges' of the paper are nicely preserved. Figure 6.12c shows the 3D data of a single line profile from one acquired frame. The right hand side of the figure magnifies a region of only $\sim 6 \times 6\ \text{mm}^2$. It can be seen that 3D data are acquired close up to the edge. Artifacts, caused by smoothing or correlation are not visible.

Although the human face is a relatively smooth object, it nevertheless contains regions with high object frequencies. Especially within the range of the eye-socket or mouth, existing single-shot approaches frequently encounter problems as shown e.g. in the section about Microsoft's Kinect-sensor (see Fig. 3.5c). Figure 6.13 shows another example containing raw data of the discussed regions, also taken from a 3D

[4] Alternative link: tinyurl.com/3DCam-007.

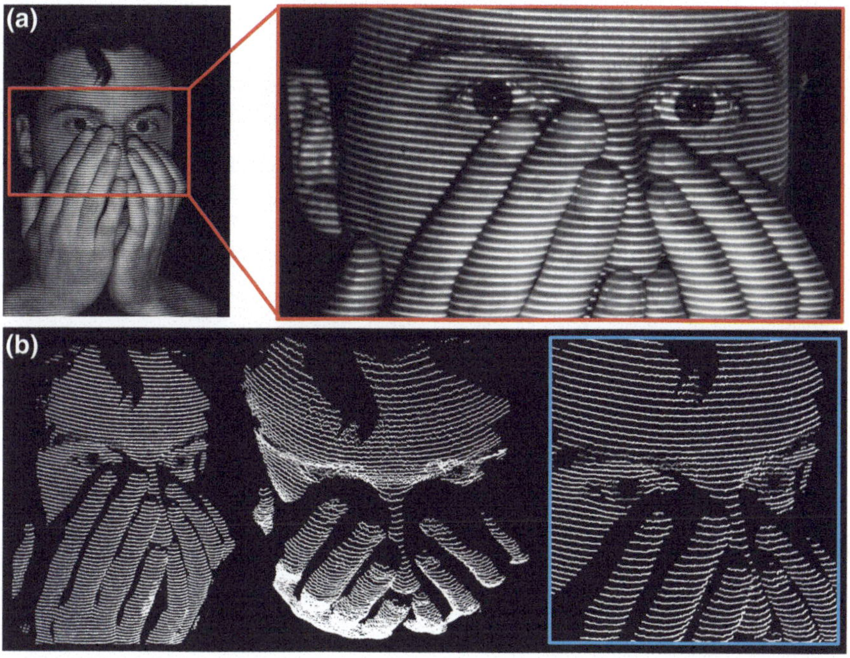

Fig. 6.13 3D measurement of a human face containing high object frequencies. The person puts his hands in front of the face. **a** Camera image (C_2) with magnification window. **b** 3D raw data from three different perspectives (including close-up view). All high object frequencies are resolved

movie sequence [9, 10] (direct link: tinyurl.com/3DMovCam04).[5] The person puts his hands in front of the face, which creates additional high object frequencies at the 'jump edges' at the fingers. All high object frequencies are not 'smoothed out' in the 3D data set. More 3D models, also containing close-up views, are shown in the subsequent chapters.

6.4 Evaluation Method and Angle Configuration

The introduced *index-back-projection-approach* that enables the *single-shot 3D movie camera*, is mainly based on the combination of two ideas:

- Arrangement of two cameras with a large and a small triangulation angle.
- Exchange of information between the two cameras via back projection over 3D space.

One idea is purely hardware-based, while the other idea is based on algorithms. Their *combination* enables the special feature of the 'index back projection', which

[5] Alternative link: tinyurl.com/3DCam-004.

manifests in the evaluation strategy of the *index-back-projection-approach* (see Sect. 5.2.1).

This section discusses, whether this *evaluation strategy is 'optimal'*, both from a software-sided and hardware-sided point of view. The related questions are:

- Is the proposed angle combination of a small and a large triangulation angle really optimal or would another combination yield better results?
- Is the proposed evaluation method of 'index back projection' really optimal or would another method yield better results?

First, the term 'better results' has to be clarified. The measurement uncertainty and data density of the prototype sensor have been already considered. It has been discussed that the triangulation angle θ_2 is already at its limit. It cannot be increased for high surface inclinations without decreasing the projected line number N_L (or vice versa). $\theta_2 = 9°$ and $N_L = 160$ are therefore considered to be ***fixed*** for the following discussions.

A quantity, which was not discussed in detail yet is the *measurement range*. By definition, an object surface, located within the *effective measurement range* ΔZ_e can be measured without ambiguities. Hereinafter 'better results' are referred to as 'larger effective measurement range ΔZ_e'.

6.4.1 Effective Measurement Range for the Index-Back-Projection-Approach Compared to an Alternative Method

In the *index-back-projection-approach*, the *unique measurement range* ΔZ_1 of the first sub-sensor defines the upper bound for the usable *effective measurement range* ΔZ_e. If a larger effective measurement range is desired (and if N_L should remain constant), the triangulation angle θ_1 has to be decreased according to Eq. (4.4). Note that an increase of ΔZ_1 is counterproductive in the current prototype setup due to the limited depth of field of the projector. However, other setups with projections having larger depths of field may profit from an extension of ΔZ_1. Hence, one could ask for the *minimal possible* θ_1.

In accordance with the findings made in Sect. 6.2.3, the 'scattering' of the back projections on the chip of C_2 has to be smaller than one line distance d'_c. In the prior section, θ_2 was determined in that way, that the line distance on the chip of C_2 will not get smaller than $d'_c = 3\ pix$. With Eq. (6.12), the minimal possible θ_1 can be therefore calculated to

$$\theta_1 = \arcsin\left(\frac{2\,\delta x'}{d'_c} \cdot \sin\theta_2\right) \approx \arcsin\left(\frac{2 \cdot 0.1\ pix}{3\ pix} \cdot \sin 9°\right) \approx 0.6°\,, \qquad (6.13)$$

with $\delta x' \approx 0.1\ pix$ being the 'worst' measured lateral localization uncertainty. According to Eq. (4.4), the related *theoretical* unique measurement range is calculated to

$$\Delta Z_1 = \frac{\Delta X}{N_L \cdot \tan \theta_1} \approx 180\ \text{mm}\ . \tag{6.14}$$

Note: These specifications cannot be realized in practice. The considerations are purely theoretical and assume a *perfect* calibration as well as no further imperfections of the setup.

In Sect. 5.2.1, the *direct comparison of 3D points in space* was briefly discussed as an alternative to the index back projection. At a first glance, this approach seems to be very similar, albeit connected to larger computational effort: For each surface point to be measured, the two triangulation sensors perform an independent 3D evaluation. A unique 3D point with poor precision is evaluated from C_1, whereas many ambiguous 3D points with good precision are evaluated from C_2. Eventually, the points are compared in 3D space. That point from C_2, overlapping with the point from C_1 *within its uncertainty interval* $2\delta z_1$, is considered to be correct.

This method works as long as the *uncertainty interval of the first sensor is smaller than the vertical (z-) distance of two ambiguous points from the second sensor* which is exactly its unique measurement range ΔZ_2.

$$2\,\delta z_1 = \frac{2\,\delta x}{\sin \theta_1} \overset{!}{<} \frac{\Delta X}{N_L \cdot \tan \theta_2} = \Delta Z_2 \tag{6.15}$$

The minimal possible triangulation angle θ_1 is therefore calculated to

$$\theta_1 = \arcsin\left(\frac{2\,\delta x}{\Delta X} \cdot N_L \cdot \tan \theta_2\right) \approx 0.3^\circ\ , \tag{6.16}$$

leading to a nearly doubled theoretical unique measurement range of

$$\Delta Z_1 = \frac{\Delta X}{N_L \cdot \tan \theta_1} \approx 355\ \text{mm}\ . \tag{6.17}$$

The resulting difference in the minimal θ_1 for both approaches can be explained as follows: The *direct comparison in 3D space* does not rely on any back projection, meaning that no 'scattering condition' (Eq. (6.12)) must be satisfied. If θ_2 is fixed, the only restriction on θ_1 is the noise level of the first sensor, which is (geometrically) independent of object inclinations α.

For smaller object inclinations, d_c' in Eq. (6.13) is defined by Eq. (6.2): d_c' increases with decreasing α and the minimal θ_1 for the index-back-projection-approach converges towards the value, calculated for the direct 3D comparison in Eq. (6.17). For $\alpha = 0^\circ$, Eq. (6.13) would also yield to $\theta_1 \approx 0.3^\circ$. However, since higher surface inclinations should not be excluded, the first answer to the posed questions is:

The *index-back-projection-approach* is ***not optimal*** in terms of a largest possible effective measurement range, since the 'scattering condition' of Eq. (6.12) introduces an additional limit to θ_1.

Nevertheless, this thesis further pursues the index-back-projection-approach due to its algorithmic simplicity. Methods, based on a direct comparison of 3D points in space, that intend to further increase the unique measurement range are discussed in Chap. 10.

6.4.2 Effective Measurement Range for the 'Large-Small' Angle Configuration Compared to Alternative Combinations

The combination of two cameras, comprising a large and a small triangulation angle with the projection, was introduced to overcome the ambiguity problem expressed in Eq. (4.4). This enables the possibility to acquire unique 3D data within an effective measurement range $\Delta Z_e = \Delta Z_1$ considerably *larger* than the unique measurement range ΔZ_2.

In analogy to the previous section, one could ask whether the proposed *angle configuration* really achieves *the largest possible effective measurement range* ΔZ_e. In the following an alternative approach, inspired by *multi-frequency phase shifting*, is used as a comparison.

In *phase measuring triangulation* (as well as many other time sequential metrology principles, e.g. *phase shifting interferometry*) ambiguities can be resolved by *multi-frequency phase shifting*. Although this method comes in different variations (see e.g. [11–13]), the basic idea can be explained as follows: After the first acquired image sequence, with each image relating to a phase shift of the projected sinusoidal pattern with frequency ν_1, a second image sequence is acquired. This second sequence comprises images related to phase shifts of a second sinusoidal pattern with frequency ν_2. After the evaluation of the phase in each sequence, each pixel (x_i', y_i') on the camera chip corresponds to *two* phase values $(\varphi_i(\nu_1), \varphi_i(\nu_2))$. Per definition, the combination $(\varphi_i(\nu_1), \varphi_i(\nu_2))$ is *unambiguous* in the whole effective measurement range. For higher ambiguity levels, additional projected patterns with different frequencies can be applied.

The basic idea of multi-frequency phase shifting can be 'translated' into single-shot approaches. In the proposed method (from now on: '*nonius-method*'), the role of the two different frequencies ν_1, ν_2 is adopted by the two different unique measurement ranges $\Delta Z_1, \Delta Z_2$. The method is explained by the simplified example of Fig. 6.14. The example is dimensionless and assumes telecentricity.

The two examined sensors comprise unique measurement ranges of $\Delta Z_1 = 4$ (blue) and $\Delta Z_2 = 3$ (yellow), which results in an effective measurement range of $\Delta Z_e = 12$. As shown in Sect. 4.2, a surface point, measured outside the unique measurement range, is always assigned to a value inside the unique measurement

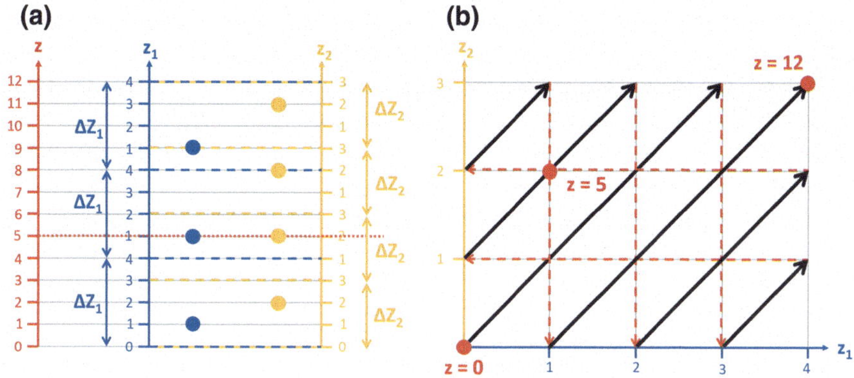

Fig. 6.14 Alternative approach for a large effective measurement range, applying two cameras. **a** $\Delta Z_1 = 4$ and $\Delta Z_2 = 3$ results in $\Delta Z_e = 12$. **b** Different representation: All possible combinations (z_1, z_2) are located on the black arrows in the diagram. The combinations are unique provided that all black arrows in the diagram are separated

range. In Fig. 6.14 it is assumed that the true depth value is $z = 5$. Due to Fig. 6.14a, $z = 5$ would be evaluated from the first sensor as $z_1 = 1$ and from the second sensor as $z_2 = 2$. The evaluated combination $(z_1, z_2) = (1, 2)$ is unique in the whole effective measurement range and can therefore be assigned to the true value $z = 5$.

The diagram of Fig. 6.14b shows an alternative schematic representation of the discussed example. The maximal value on the horizontal axis is $\Delta Z_1 = 4$, whereas the maximal value on vertical axis is $\Delta Z_2 = 3$. All possible combinations (z_1, z_2) are located on the black arrows. The combinations are unique provided that all black arrows in the diagram are *separated* and *do not overlap*. The effective measurement range is the sum of the length of all arrows divided by $\sqrt{2}$.

So much for the introducing example. A diagram as seen in Fig. 6.14b can be drawn for many possible combinations of ΔZ_1 and ΔZ_2. Without loss of generality, it is assumed for the following that $\Delta Z_1 > \Delta Z_2$, meaning respectively that $\theta_1 < \theta_2$ and $\delta z_1 > \delta z_2$.

In general, the effective measurement volume ΔZ_e created by the above described *'nonius method'* can be expressed by[6]

$$\Delta Z_e = \frac{\Delta Z_1 \cdot \Delta Z_2}{\Delta Z_1 - \Delta Z_2} . \tag{6.18}$$

In accordance with the previously discussed requirements for high precision, lateral resolution and data density, the values of $\theta_2 = 9°$, $N_L = 160$, $\Delta X = 300$ mm, and

[6]It is striking, that Eq. (6.18) is very well known in multi-wavelenghts Interferometry [12, 14], where exactly the same principle is exploited. In ordinary Interferometry, the unique measurement range of the sensor is defined by the used wavelength λ. By using two different wavelengths λ_1 and λ_2, the unique measurement range can be extended to $\Lambda_e = \frac{\lambda_1 \cdot \lambda_2}{|\lambda_1 - \lambda_2|}$. Λ_e is often referred as 'synthetic wavelength' or 'effective wavelength' and represents the beat wave between λ_1 and λ_2.

$\delta x \approx 31.3\,\mu\text{m}$ remain *fixed*, which means that ΔZ_2 is also fixed and could be directly calculated. In this case the largest effective measurement range ΔZ_e can be achieved if the difference between ΔZ_1 and ΔZ_2 is as small as possible. In the scheme of Fig. 6.14b, this can be interpreted as the *densest possible packing of black arrows*. However, the density of packing, i.e. the minimal detectable difference between ΔZ_1 and ΔZ_2, is limited by the noise on the data. Figure 6.15a visualizes the *maximal allowed noise level* for the introduced example: For a unique separation of the black arrows their horizontal distance (which equals to their vertical distance) has to be larger than the highest noise interval in the system, i.e. $2\delta z_1$. This means that the minimal allowed difference between ΔZ_1 and ΔZ_2 is $\Delta Z_1 - \Delta Z_2 = 2\delta z_1$ (see Fig. 6.15b). From Eqs. (2.20) and (4.4), the relation[7] between ΔZ_1 and δz_1 is found to be

$$\frac{\Delta Z_1}{\delta z_1} = \frac{\Delta X}{\delta x \cdot N_L} \,. \tag{6.19}$$

This means that the maximal possible effective measurement range for the prototype setup using the alternative indexing approach of the *'nonius-method'* can be finally calculated to

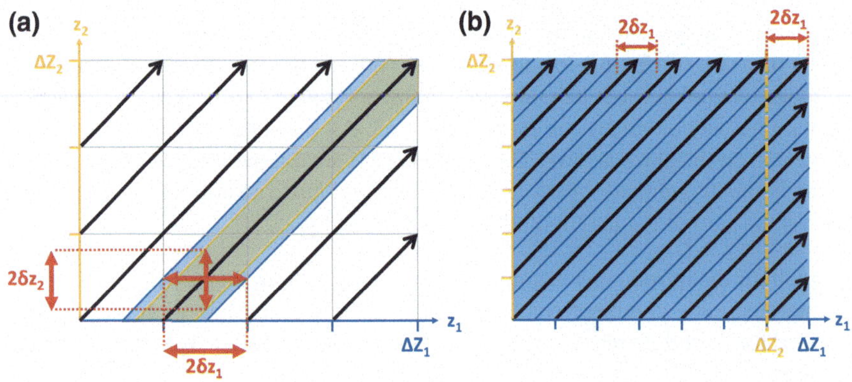

Fig. 6.15 Theoretical optimization of the alternative approach. **a** Maximal allowed noise level for the introduced example of Fig. 6.14. **b** Condition for a largest possible effective measurement range

[7]Note that the strict application of Eqs. (2.20) and (4.4) would lead to the expression $\delta z_1 = \frac{\delta x}{\sin\left(\arctan\left(\frac{\Delta X}{\Delta Z_1 \cdot N_L}\right)\right)}$. The reason for this is that the well known formula for distance uncertainty in triangulation (Eq. (2.20)) and the formula in Eq. (4.4) have been derived under slightly different geometrical assumptions (compare [15] and [16, 17]): For Eq. (2.20), the viewpoint on the scenery is defined by the *projection* [15], whereas for Eq. (4.4) the viewpoint is defined by the *camera* [16, 17]. This difference in definition is usually ignored for most triangulation setups, since $\sin\theta \approx \tan\theta$ applies for practical (i.e. relatively small) triangulation angles anyway. If equal definitions of the viewpoints are used, Eq. (6.19) is always exact (also for larger angles).

$$\Delta Z_e = \frac{\Delta Z_1 \cdot \Delta Z_2}{2\delta z_1} = \frac{\Delta Z_2 \cdot \Delta X}{2\ \delta x \cdot N_L} = \frac{\Delta X^2}{2\ \delta x \cdot \tan\theta_2 \cdot {N_L}^2} = \tag{6.20}$$
$$= \frac{(300\ \text{mm})^2}{2 \cdot 0.0313\ \text{mm} \cdot \tan 9° \cdot 160^2} \approx 355\ \text{mm}$$

This is the same value which was calculated in Eq. (6.17). Hence, the comparison between the angle configuration used in the *single-shot 3D movie camera* ('large-small' angle configuration) and other possible angle combinations, yields to the following conclusion:

> The *'large-small' angle configuration* used in the *single-shot 3D movie camera* ***is optimal!*** All other possible angle combinations, derived from a method for single-shot principles inspired by multi-frequency phase shifting ('nonius-method'), *yield the same theoretical maximum* for the effective measurement range of $\Delta Z_e \approx 355$ mm for the example of the prototype setup.

Again, such a large ΔZ_e can not be reached in practice, due to imperfections of the setup and the calibration. Moreover, the situation is different (and much more complicated), if the non-telecentric case is considered.

6.5 Resulting Channel Efficiency

After the discussions of the previous sections, the *channel efficiency* of the single-shot 3D movie camera can be finally estimated. This requires the quantification of all variables in Eq. (2.32) as follows:

- F and Q: The sensor is *single-shot*. It requires only $F = 1$ frame with $Q = 2$ simultaneously acquired images.

- ρ_{3D}: The illumination projects up to 160 lines. Due to slightly different aspect ratios and field sizes of projector and cameras, not all lines are always visible in the camera images. For the following calculations the realistic number of $N_L = 150$ 'visible' lines is considered. 3D points, independent of the other 3D points, are evaluated from each row of each (vertical) line in the camera image. The resulting 3D data density for the prototype setup is

$$\rho_{3D} = \frac{N_{3D}}{N_{chip}} = \frac{N_L}{N_x} = \frac{150}{1024} = 14.6\,\% \,. \tag{6.21}$$

- $\Delta Z/\delta z$: Section 6.2 revealed that the measurement uncertainty is dependent on the properties of the measured surface, and is influenced by technical properties like aberrations. The resulting *dynamical range* of the sensor, defined by $\Delta Z/\delta z$, varies from **500** up to **5800** at best.[8] For the calculation of the channel efficiency, the

[8] Best measured values from the evaluation on the calibration plate (Fig. 6.7): $\delta z_{min}(z = 470\ \text{mm}) = \delta z_{min}(z = 480\ \text{mm}) = 17\ \mu\text{m}$. The resulting *dynamical range* is $\Delta Z/\delta z = 5882$.

pessimistic value of **500** is assumed, which corresponds to $\Delta Z = \Delta Z_1 = 100$ mm, divided by the worst evaluated value $\delta z_{max} = 200\ \mu$m.

- $\underline{SNR}$: As discussed in Sect. 2.3.3 the value of $\log_2(1 + SNR)$ is estimated to 6 bit. This value has been empirically confirmed several times for different triangulation sensor setups and is considered as 'good estimate'.

Using these values, the *channel efficiency* of the single-shot 3D movie camera prototype is calculated to

$$\eta_{3DC} = \frac{\rho_{3D}}{Q\ F} \frac{\log_2(1 + \frac{\Delta Z}{\delta z})}{\log_2(1 + SNR)} = \frac{0.146}{2 \cdot 1} \frac{\log_2(1 + \frac{100\ \text{mm}}{0.2\ \text{mm}})}{6\ bit} \approx 11\% \ . \tag{6.22}$$

For a better judgment of this result, the channel efficiency of the single-shot 3D movie camera is compared with three other channel efficiencies of established triangulation approaches: Common *multi-line triangulation*, *active stereo* and *phase measuring triangulation*. For the sake of comparison, the channel efficiency of the other approaches is calculated for a setup with the same parameters as the prototype setup. Consequentially, some of the considered values are only rough estimations, in accordance with the performance of real setups with different dimensions.

As a representative approach for *precise (incoherent) multi-line triangulation*, Flying Triangulation is considered. As Flying Triangulation is also 'physically optimized' to a large extent, it is assumed to display the same dynamical range as the 3D movie camera prototype for a setup with the same dimensions. However, only $Q = 1$ camera image is used and the 3D data density per single-shot is much lower. Due to Eq. (4.4), the dimensions of the prototype setup would allow for the projection of $N_L = 19$ vertical lines. The resulting channel efficiency is therefore:

$$\eta_{FlyTri} = \frac{\rho_{3D}}{Q\ F} \frac{\log_2(1 + \frac{\Delta Z}{\delta z})}{\log_2(1 + SNR)} = \frac{19/1024}{1 \cdot 1} \frac{\log_2(1 + \frac{100\ \text{mm}}{0.2\ \text{mm}})}{6\ bit} \approx 2.8\% \tag{6.23}$$

To calculate the channel efficiency of *active stereo* approaches, a method, similar to the method that is used in Microsoft's Kinect One® with $Q = 1$ camera, and a calibrated projection (see Sect. 3.3.1) is considered. For an unambiguous correlation of the projected binary pattern, a typical correlation window size of $7 \times 7\ pix$ is assumed [18]. This relates to a data density of $\rho_{3D} = 1/49$. For incoherent illumination (which is not the case for the MS Kinect®!), the above used value of $\log_2(1 + SNR) = 6\ bit$ can also be reached for active stereo. Under this condition, the correlation of the pattern in the camera images can be performed with a lateral precision up to $\delta x' \approx 0.1\ pix$ [19]. For the same triangulation angle ($\theta = 9°$) and dimensions of the measurement volume as the prototype setup, this leads to a distance precision not better than $\delta z \approx 200\ \mu$m. The channel efficiency of comparable optimized active stereo approaches is therefore estimated to:

$$\eta_{AS} = \frac{\rho_{3D}}{Q\ F} \frac{\log_2(1 + \frac{\Delta Z}{\delta z})}{\log_2(1 + SNR)} = \frac{1/49}{1 \cdot 1} \frac{\log_2(1 + \frac{100\ \text{mm}}{0.2\ \text{mm}})}{6\ bit} \approx 3\% \tag{6.24}$$

Phase measuring triangulation (PMT) relies on subsequently acquired images. For a setup with comparable properties, a required image sequence of $F = 8$ images is assumed, which corresponds to a four-phase shift with two different frequencies. The acquired 3D data density is approximated to be $\rho_{3D} = 100\%$. In optimized setups, the fringe phase can be evaluated with a very high lateral precision of $\delta x \approx 1/40\ pix$, leading to a distance precision of $\delta z \approx 50\ \mu$m for a setup with the same measurement volume as the prototype. Although such a high distance precision requires commonly a very good SNR in the 2D images, $\log_2(1 + SNR) = 6\ bit$ is also assumed here for a 'fair' comparison (note that the channel efficiency would be *reduced* if a higher value would be assumed). The resulting channel efficiency calculates to:

$$\eta_{PMT} = \frac{\rho_{3D}}{Q\ F} \frac{\log_2(1 + \frac{\Delta Z}{\delta z})}{\log_2(1 + SNR)} = \frac{1}{1 \cdot 8} \frac{\log_2(1 + \frac{100\ \text{mm}}{0.05\ \text{mm}})}{6\ bit} \approx 22.8\% \qquad (6.25)$$

Note: The acquisition of 3D data with $\delta z = 50\ \mu$m precision, comprising only 8 subsequent images, is a very optimistic scenario. High precision PMT setups rely on the acquisition of sequences up to $F = 30$ images (or even more). The channel efficiency decreases inversely to F. For, e.g., $F = 16$ images the channel efficiency would reach a value of $\eta_{PMT} \approx 11.4\%$, which is comparable to the value for the single-shot 3D movie camera.

Comparing the estimated channel efficiencies of the considered principles with the 3D movie camera, it can be said that:

> The prototype of the *single-shot 3D movie camera* in its current form displays a channel efficiency which is *significantly higher than the channel efficiencies of established single-shot approaches*. Moreover, the determined value is *in the range of good PMT approaches*, albeit the high channel efficiency of very good PMT approaches can not be reached so far.

Although PMT relies on a temporal image sequence, it is obviously not considered as *'the gold standard of optical 3D metrology'* for nothing. Very good setups are able to reach channel efficiencies close to the estimated information theoretical maximum of $\eta \leq 33\%$.

Nevertheless, the next chapter reveals that the channel efficiency of the 3D movie camera *can still be improved*! Although the minimal possible distance of projected lines is already reached, the *3D data density* of the 3D movie camera can be ***nearly doubled*** by a simple trick. By applying this trick, the single-shot 3D movie camera prototype displays a channel efficiency of $\eta_{3DC+} \approx 20\%$, which is finally in the range of very good PMT approaches.

References

1. O. Arold, Flying Triangulation-handgeführte optische 3D Messung in Echtzeit. Dissertation, University Erlangen-Nuremberg (2013)
2. M. Knauer, Signalverarbeitung und neue Anwendungsgebiete des Spektralradars. Diploma Thesis, University Erlangen-Nuremberg (1999)
3. P. Vogt, Beleuchtungsoptimierung für einen bewegungsunempfindlichen 3D-Sensor. Diploma Thesis, University Erlangen-Nuremberg (2008)
4. Z. Yang, Ein miniaturisierter bewegungsunempfindlicher 3D-Sensor. Master Thesis, University Erlangen-Nuremberg (2009)
5. Acer, Projector K132 Datasheet (2017). www.tinyurl.com/AcerK132-Details-pdf. Accessed 15 May 2017
6. F. Schiffers, Kalibrierung von Multilinientriangulationssensoren. Bachelor Thesis, University Erlangen-Nuremberg (2014)
7. X. Laboureux, G. Häusler, Localization and registration of three-dimensional objects in space—where are the limits? Appl. Opt. **40**(29), 5206–5216 (2001)
8. B. Curless, M. Levoy, Better optical triangulation through spacetime analysis. in *Proceedings of the Fifth International Conference on Computer Vision* (1995), p. 987
9. Video file repository for Florian Willomitzer's Dissertation (urn:nbn:de:bvb:29-opus4-85442). University Erlangen-Nuremberg. nbn-resolving.de/urn:nbn:de:bvb:29-opus4-85442. Accessed 31 May 2017
10. Osmin3D, YouTube-channel of the OSMIN research group at the University Erlangen-Nuremberg. www.youtube.com/user/Osmin3D. Accessed 15 May 2017
11. H.O. Saldner, J.M. Huntley, Temporal phase unwrapping: application to surface profiling of discontinuous objects. Appl. Opt. **36**(13), 2770–2775 (1997)
12. Y. Cheng, J. Wyant, Two-wavelength phase shifting interferometry. Appl. Opt. **23**(24), 4539–4543 (1984)
13. K. Falaggis, D.P. Towers, C.E. Towers, Method of excess fractions with application to absolute distance metrology: theoretical analysis. Appl. Opt. **50**(28), 5484–5498 (2011)
14. C. Polhemus, Two-wavelength Interferometry. Appl. Opt. **12**(9), 2071–2074 (1973)
15. R.D. Dorsch, G. Häusler, J.M. Herrmann, Laser triangulation: fundamental uncertainty in distance measurement. Appl. Opt. **33**(7), 1306–1314 (1994)
16. G. Häusler, W. Heckel, Light sectioning with large depth and high resolution. Appl. Opt. **27**(24), 5165–5169 (1988)
17. F. Willomitzer, S. Ettl, C. Faber, G. Häusler, Single-shot three-dimensional sensing with improved data density. Appl. Opt. **54**(3), 408–417 (2015)
18. C. Wagner, Informationstheoretische Grenzen optischer 3D-Sensoren. Dissertation, University Erlangen-Nuremberg (2003)
19. C. Perwaß, L. Wietzke, Single lens 3D-camera with extended depth-of-field. IS&T/SPIE electronic imaging. in *International Society for Optics and Photonics* (2012)

Chapter 7
Further Improvements of the Single-Shot 3D Movie Camera

This chapter introduces two methods, which significantly improve the measurement outcome of the single-shot 3D movie camera:

- The further increase of the 3D data density *and* channel efficiency by the application of a projected pattern with *crossed lines*.
- The acquisition of *texture information*, which enhances the visual impression of the acquired 3D model and could also serve as an additional source of information.

Both methods require a slight softening of the 'pure doctrine' of the previous two Chaps. 5 and 6, as now additional space-bandwidth restrictions are introduced. In exchange, both improvements are applicable without the need of additional cameras in the setup.

7.1 With Crossed Lines Towards Higher Data Density

The discussion in Sect. 6.1 revealed that the line density, projected by the 3D movie camera prototype, is already close to its upper limit for the applied triangulation angle of $\theta_2 = 9°$. More precise: The *horizontal distance* between two projected vertical lines can not decrease further. Nevertheless, there is a possibility to significantly increase the 3D data density: The projection of a *pattern with crossed lines*.

This means, that in addition to the projected vertical lines, a multitude of horizontal lines is projected simultaneously. Since both line directions are independent from each other in terms of their maximal density, the horizontal lines can be projected with the same period as the vertical lines. The original picture projected by the illumination unit is shown in Fig. 7.1. According to the 16 : 10 aspect ratio of the projector it consists of $N_{L,v} = 160$ vertical and $N_{L,h} = 100$ horizontal lines (the reader might zoom in to resolve all lines). For comparison, the projected unidirectional pattern with $N_{L,v} = 160$ vertical lines, is shown in Fig. 6.3a.

F. Willomitzer, *Single-Shot 3D Sensing Close to Physical Limits and Information Limits*, Springer Theses,
https://doi.org/10.1007/978-3-030-10904-2_7

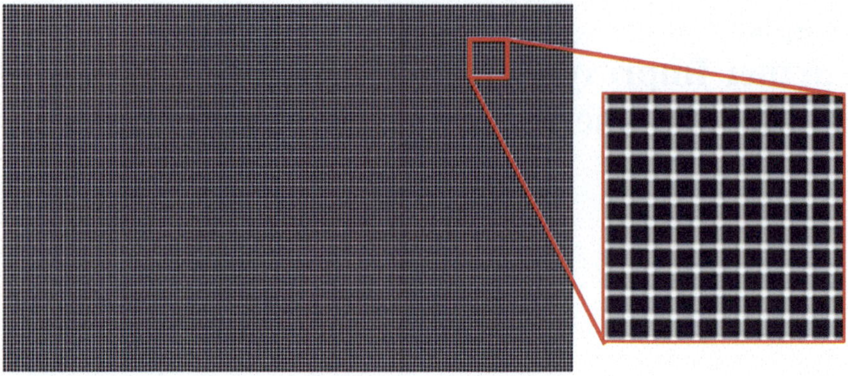

Fig. 7.1 Projected pattern with crossed lines

For the generation of 3D data with such a crossed pattern, the vertical and horizontal lines are *separated* in each acquired camera image and *evaluated independently*. This requires additional modifications for the prototype setup and the evaluation algorithms. The three major questions to be solved are:

- How to separate the two line directions?
- How to triangulate the ('new') horizontal line direction?
- What happens at the crossing points?

The answers are given in the following subsections.

7.1.1 Separation of the Two Line Directions

There are several options for the separation of lines with two different directions. The easiest would be differentiation in each direction. It turns out that this is sensitive to different acquisition scenarios and requires a precise adjustment of parameters for each measurement.

A more robust method is the filtering of the line directions in the *Fourier-domain*: Each acquired camera image is Fourier-transformed and the high frequency components of one direction is masked. The back transformed image contains only lines in one direction.

Note that the lines in the acquired camera images are *deformed*, i.e. not perfectly vertical and horizontal, due to the height variations of the object surface. This means a low pass is also introduced on the lines which are not filtered. In order to avoid low pass filtered 3D data, the Fourier-filtered image is *not* used for the data evaluation! Instead, the filtered image is applied as a *mask* on the originally acquired image, which is then used in the further evaluation process. This preserves the resolution of small object features along a projected line.

Nevertheless, the filtering has a certain price: If a signal of a line is split into many separated points (due to dynamic surface variations), this line would not pass the filtering process. At least a small line segment has to be visible in the camera image, which requires a few neighboring line points. The result is a (not too serious) bandwidth restriction of the object surface.

In Chap. 8 the separation of the two line directions will be illustrated for a concrete evaluation example.

7.1.2 Crossing Points

During the three-dimensional evaluation of a (vertical) line profile, the sub pixel precise intensity maximum of the line is evaluated in each pixel-row of the line via Gauß-approximation (for horizontal lines in each pixel-column). At the crossing points, the line profile displays no intensity variation, even if the other line direction is masked out. The evaluation of a Gauß maximum and hence the generation of a precise 3D point is impossible. This case is schematically illustrated in Fig. 7.2a.

To generate 3D points at the crossing points anyway, a simple trick is applied: The intensity of the projected lines is slightly reduced except for the projector pixels[1] at the crossing points (see Fig. 7.2b). By this method, a Gauß maximum can be evaluated in *each* pixel row (or column) of the line. The price is a reduced SNR, which leads to a lower precision of 3D data. In practice, an intensity ratio of $1/0.7$ between crossing point and line has been proven beneficial.

Another possibility for line maxima evaluation at the crossing points is the Gauß evaluation *along a diagonal pixel array* (see Fig. 7.2c). Here the projected intensity remains constant. However, the diagonal profile at the crossing point is wider, which

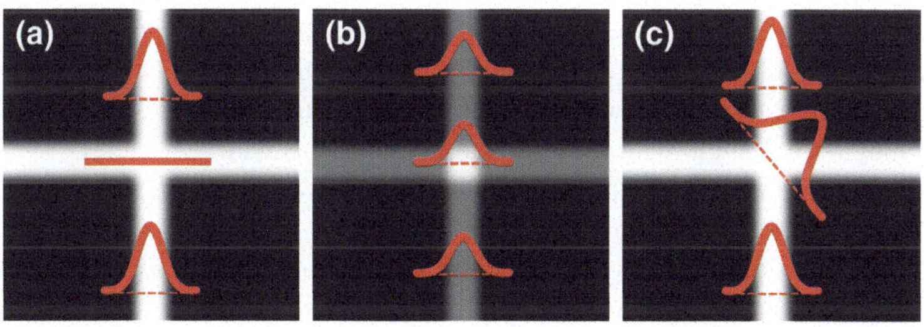

Fig. 7.2 Gauß approximation at the crossing points. **a** Row-wise sampling, no intensity variation. Precise evaluation at the crossing point is not possible. **b** Row-wise sampling with intensity variation. Evaluation is possible, SNR and precision are reduced. **c** Diagonal sampling, no intensity variation. Profile at the crossing point is wider, for this point the precision is reduced

[1] This could also be realized for a slide-projector setup comprising a chrome-on-glass mask. Varying intensity is reached by different line widths or different dithering of the lines on the mask. See [1, 2] for further information.

also leads to a reduced localization precision of the maximum. Moreover, the method requires additional computational effort, since the algorithm has to 'know' the exact positions of the crossing point-pixels and the evaluation procedure has to be modified at these points.

In the current version of the single-shot 3D movie camera the method visualized in Fig. 7.2b is used due to its 'algorithmic simplicity', which requires no further modification of the evaluation algorithm.

7.1.3 Adapted and Optimized Sensor Setup

In order to triangulate the horizontal line direction in the crossed pattern, the horizontal lines have to enclose a certain triangulation angle with each camera. In most line triangulation setups (e.g. the schematic explanation of Fig. 3.7), camera(s) and projector are aligned along the triangulation base and the lines are projected perpendicularly to this base. If a crossed pattern would be projected, the second line direction would enclose a triangulation angle $\theta = 0°$ with the camera(s), making a triangulation impossible.

To triangulate the horizontal lines in the setup of the single-shot 3D movie camera, principally a *second pair of cameras* can be added, aligned along a second (perpendicular) triangulation base with the projector. However, such an approach is not favorable. It requires additional hardware as well as additional images for the evaluation. Although the *3D data density* of the sensor is improved, the *channel efficiency* would not increase!

However, there is an alternative solution *requiring no additional hardware*: As discussed in Sect. 5.2.1, only the distance of projector and camera *perpendicular to the line direction* contribute to the triangulation angle. This fact was already exploited in the prototype setup of Fig. 5.14, since the bulky projector did not allow for a linear alignment. Now, it can be used to *save cameras and channel capacity*. The two cameras of the prototype setup are realigned in a way that *each camera* -C_1 and C_2- comprises a large triangulation angle with one line direction and *simultaneously* a small triangulation angle with the other line direction as well.

The resulting setup of the prototype sensor is shown in Fig. 7.3. As before, the *vertical* lines comprise a small triangulation angle $\theta_{1,v} = 1°$ with camera C_1 and a large triangulation angle $\theta_{2,v} = 9°$ with camera C_2. In contrary, the *horizontal* lines comprise a large triangulation angle $\theta_{1,h} = 9°$ with camera C_1 and a small triangulation angle $\theta_{2,h} = 1°$ with camera C_2.

This means that *the role of both cameras is changed for the evaluation of the horizontal lines*: A noisy, but unique 3D model is evaluated from the image of C_2 and back projected onto the chip of C_1 with the index being asserted and the final 3D model being calculated.

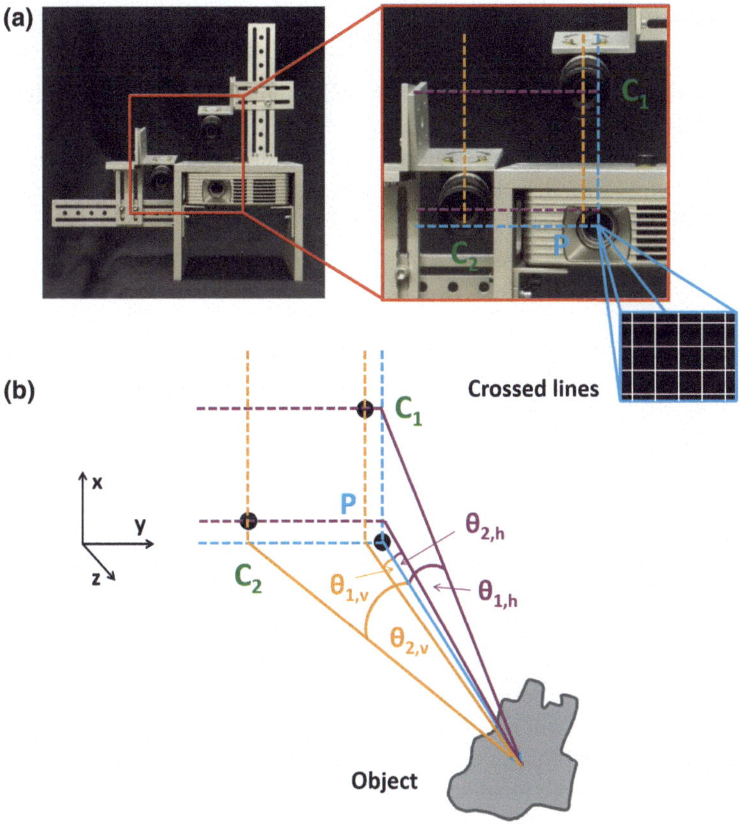

Fig. 7.3 Modified prototype setup for the 3D evaluation of crossed lines. The two cameras are aligned in a way, that they simultaneously span a large and a small triangulation angle with each line direction. **a** Setup with marked lateral distances between cameras and projector. **b** Schematic drawing with related angles

The modified prototype setup displays a remarkable flexibility. Since the switchable LED projector is still part of the setup[2] either *only vertical*, or *only horizontal*, or *crossed* lines can be evaluated without any hardware changes. If a slide-projector is applied for illumination, only the chrome-on-glass mask has to be replaced. Due to the reverse-symmetric angle configuration, each line direction produces 3D data with the same measurement uncertainty. Asymmetric angle configurations are possible, too. Related prospects are discussed in Sect. 10.2.

[2]For a final version of the single-shot 3D movie camera, a slide-projector with a chrome-on-glass mask is still preferable, due to the much better illumination conditions. However, a LED projector proved to be advantageous during the development phase, since line numbers, directions, thicknesses, etc. can be changed uncomplicated.

7.1.4 Results and Discussion

With a *pattern of crossed lines*, the number of 3D points per single-shot can be nearly doubled. From the maximal 160 vertical and 100 horizontal projected lines, roughly $N_{L,v} = 150$ vertical and $N_{L,h} = 100$ horizontal lines are visible in the camera images, due to the slightly different fields and aspect ratios. This means a 3D height map acquired within one frame of the single-shot 3D movie camera prototype contains up to

$$N_{3D} = 150 \cdot 682 + 100 \cdot 1024 - 150 \cdot 100 = 189,700 \qquad (7.1)$$

3D points, if the crossing points are only counted once. Note that these points are *widely*, but *not completely* independent from each other. Although each 3D point is still evaluated from an independent array of pixels, the prior line filtering requires some neighborhood. The resulting 3D data density of

$$\rho_{3D} = \frac{N_{3D}}{N_{chip}} = \frac{189.700}{1024 \cdot 682} = 27.2\% \qquad (7.2)$$

is therefore an upper estimation of the slightly lower real value. Since no additional cameras (and images) are required for the 3D evaluation of the crossed pattern, the *channel efficiency* is also increased. Note: The SNR in the images is slightly reduced due to the projection of patterns with brighter crossing points. This slightly affects the 3D data, too. The effect will not be considered further, since only an upper estimation is to be given at this point. For the calculation, the 'worst' value $\delta z_{max} = 200\ \mu$m, measured for the unidirectional setup in Sect. 6.2 is considered. The related upper estimation for the channel efficiency is therefore calculated to

$$\eta_{3DC+} = \frac{\rho_{3D}}{Q \cdot F} \frac{\log_2(1 + \frac{\Delta Z}{\delta z})}{\log_2(1 + SNR)} = \frac{0.272}{2 \cdot 1} \frac{\log_2(1 + \frac{100\ mm}{0.2\ mm})}{6\ bit} = 20.3\%\,. \qquad (7.3)$$

This means, that the *channel efficiency* of the 3D movie camera is now comparable to *high quality PMT setups* (see Eq. (6.25)).

Figure 7.4 shows the single-shot measurement of a human face, acquired with 160 vertical and 100 horizontal projected lines. The camera image of C_1 is shown in Fig. 7.4a. According to the angle configurations, one line direction displays a deformation (large θ), while the other direction is nearly straight (small θ) – to be seen in the magnification window. Moreover, the fine sampling of the object surface[3] with the crossed line pattern is visible. Figure 7.4b displays the 3D model from two viewpoints and a close-up view. The fine details in the area around the eyes are still preserved in the 3D dataset.

Besides the mentioned improvements, the discussion of the previous section revealed an additional insight: In fact, the modified setup of Fig. 7.3 comprises ***four*** *trian-*

[3]The distance between two lines onto the object surface is approximately 2 mm.

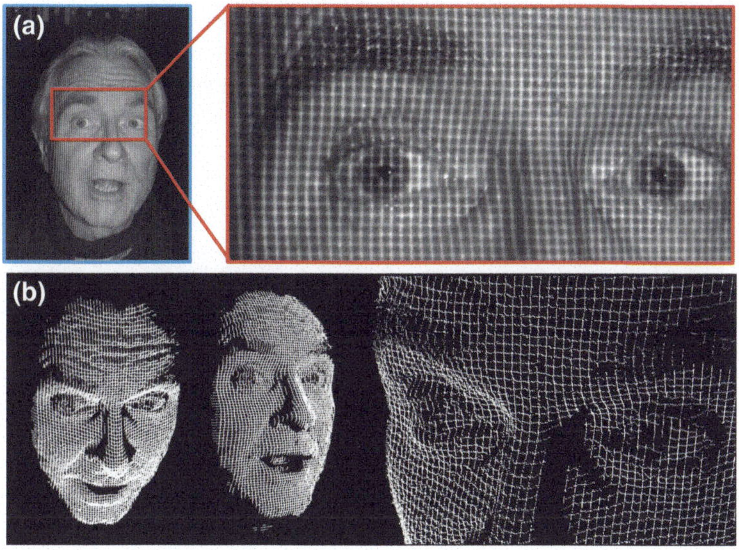

Fig. 7.4 Single-shot measurement of a human face with a projected pattern of crossed lines. **a** Original camera image of C_1. **b** Evaluated 3D model from three different viewpoints

gulation sensors! One sensor with a large and one sensor with a small triangulation angle for each line direction. These *four sensors* are created by *only two cameras*.

This principle can be *generalized*, meaning that a setup with K cameras, projecting a line grating with R directions, is able to comprise in total $K \times R$ triangulation sensors. This is not just a theoretical consideration. Principally, the crossed pattern of Fig. 7.1 can be upgraded by a *third* line direction in order to achieve an even higher data density. However, the use of this method becomes limited for additionally added line directions, since the crossing points can only be counted once. Moreover, the separation of lines enclosing an angle of 45° (or lower) becomes difficult in the camera images, since the lines are distorted.

It should be noted as well that the setup with a crossed pattern and two cameras is not restricted to the geometry shown in Fig. 7.3. One could also align both cameras along the bisectrix between x -and y -axis (or x -and $(-y)$ -axis in Fig. 7.3). One camera, say C_1, has to be aligned close to the projector, the other camera C_2 at a larger distance. This means that C_1 comprises a small angle with *both* line directions, while C_2 comprises a large angle with *both* line directions ($\theta_{1,v} = \theta_{1,h} = 1°$ and $\theta_{2,v} = \theta_{2,h} = 9°$ for the above discussed angle configuration). The evaluation sequence is adjusted accordingly.

A 'linear setup' like this yields the advantage that all components can be aligned along one axis while two line directions can still be triangulated. The form factor of the resulting sensor becomes smaller. However, this setup requires again that C_1 is spaced very close to the projection P, which might not be possible for larger objective diameters (see Sect. 5.2.1). Nevertheless, it might be the correct choice for mobile phone implementations or other applications that use objetives with a small diameter. Moreover, the 'linear setup' can be used if the *'nonius-method'*, introduced in Sect. 6.4.2 is applied for each line direction to obtain the correct index.

7.2 Texture Acquisition

The additional acquisition of texture information allows for a significant enhancement of the visual impression. Although the 3D coordinates are not changed, the acquired 3D model instantly looks more realistic. This is a well-known effect the field of computer graphics. Gaming applications are one example: In order to keep the computational effort low (so that the game runs smoothly), the 3D models in computer games only consist of a very limited number of polygons, i.e. 3D points. The visual 3D effect is mainly created by high quality color texture, which is mapped onto the 'poor' 3D model. As discussed in Chap. 3, this effect is also exploited by several 3D measurement principles to conceal the poor quality of the measurement outcome.

Commonly, the texture information is acquired with an auxiliary (color) camera, exploiting an additional exposure with flash illumination after the actual acquisition of 3D data (see e.g. [3, 4]). This procedure is *not single-shot*.

For the 3D movie camera, a different procedure was developed, which *preserves the single-shot ability*. Moreover, it neither requires additional hardware nor a change in the projected pattern. The method simply uses the *intensity along the lines*, which is reflected from the object surface. After the Gauß maximum evaluation of each line profile, the intensity value of that *pixel*, where the maximum is located, is used as the texture value of the corresponding 3D point. The image of C_1 can be used as well as the image of C_2. The difference in the reflected intensities due to the different viewing angles is considered to be negligible.

Although this kind of texture acquisition is a very rudimentary, it yields good results. Comparative examples with mapped gray-texture are displayed in Fig. 7.5.

Note that *texture* is an additional source of information. From an information theoretical point of view, this means that the *information content* of a textured 3D model is *higher* (which however is not further considered here).

Moreover, it should be clear that high quality texture acquisition and mapping is not the topic of this thesis. And indeed, the proposed method comes along with some drawbacks: First, texture information is only acquired along the projected lines. This is sufficient for the representation of 3D models via point clouds. However, for a 'watertight' representation with triangular meshes (e.g. required for 3D printing), there is no texture information in the space between the lines. The texture information has to be interpolated.

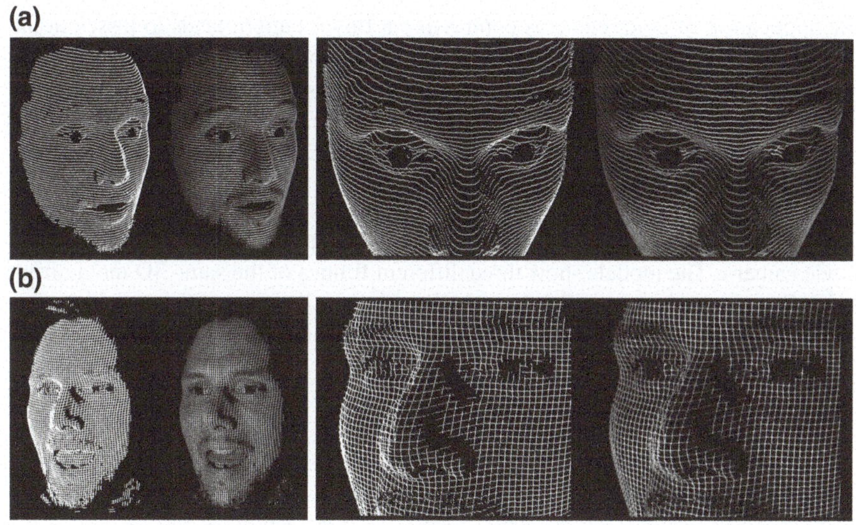

Fig. 7.5 3D models with and without mapped gray texture. Overall view and close-up view for each 3D model. **a** Acquisition with unidirectional lines. **b** Acquisition with crossed lines

Another drawback is the texture acquisition for measurements with crossed lines. Due to the higher intensity at the crossing point, the related texture value is not consistent with the neighboring values. Fortunately, this effect is practically not visible to the naked eye. A solution could be an artificial decrease of the acquired intensity values at crossing points (the *projected* intensity ratio between line and crossing point is known).

For the acquisition of *color texture* information, it is sufficient to replace *at least* one of the cameras in use by a color camera. The computational steps are analog.

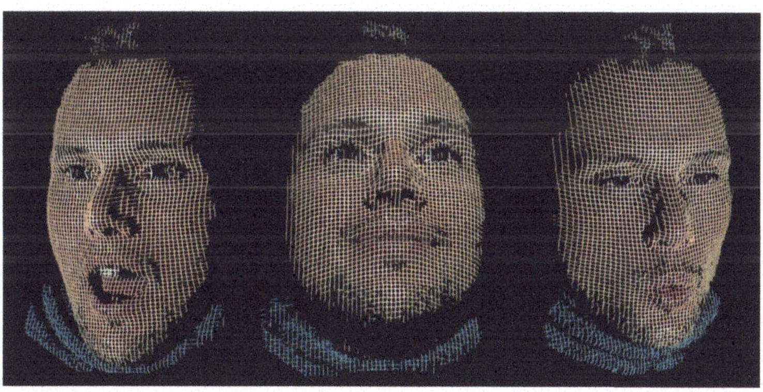

Fig. 7.6 3D models with mapped color texture acquired with a crossed line pattern. Three different frames of the same 3D movie are shown from three different perspectives

The application of color cameras comprising a Bayer-pattern leads to losses in the 3D data quality (see discussion in Sect. 3.3.5). The problem can be solved by the usage of 'three-chip-cameras'. Another (probably less expensive) solution consists of mounting (and calibrating!) an additional color camera in the setup with two monochrome cameras. The camera is not involved in 3D data acquisition and acquires only the texture along the lines for each single-shot.

Figure 7.6 displays 3D models with color texture, acquired with the single-shot 3D movie camera. The models show three different frames of the same 3D movie from three different perspectives. For the acquisition, the two monochrome cameras were simply replaced by color cameras. The resulting loss of data quality is condoned due to the algorithmic simplicity.

References

1. F. Willomitzer, FlyFace – 3D-Gesichtsvermessung mittels "Flying Triangulation". Diploma Thesis, University Erlangen-Nuremberg (2010)
2. C. Faber, New Methods and Advances in Deflectometry. Dissertation, University Erlangen-Nuremberg (2012)
3. O. Arold, S. Ettl, F. Willomitzer, and G. Häusler, Hand-guided 3D surface acquisition by combining simple light sectioning with real-time algorithms (2014), arXiv:1401.1946
4. F. Willomitzer, Z. Yang, O. Arold, S. Ettl, and G. Häusler, 3D face scanning with "Flying Triangulation". in *Proceedings of the 111th DGaO Conference* (2010), p. 18

Chapter 8
Algorithmic Implementations

This chapter explains the sequence of algorithmic steps required for the *calibration* of the single-shot 3D movie camera as well as for the *3D evaluation*. The usage of code segments or extensive mathematical explanations is largely avoided.

The *calibration* of the 3D movie camera has not been discussed so far. Section 8.1 introduces a calibration procedure which is specifically tailored to the 3D movie camera. The basic steps are explained and the advantages over existing procedures are discussed.

The explanation of the evaluation procedure is brief. The basic ideas of the evaluation steps were already outlined. Hence, Sect. 8.2 can be seen as a summary of the entire evaluation process. It is listing all required algorithmic steps in the correct order and discusses a few algorithmic approaches in more detail.

8.1 Calibration

Proper calibration plays a crucial role in the sensor concept. This is not only because each 3D point should be measured as accurate as possible. Before a final 3D point is calculated, a noisy 3D point is evaluated from one camera and back projected onto the chip of the other camera. Without a very accurate calibration of *all components* in the setup, the back projections would not overlap properly with the line images. A correct 3D evaluation would not be possible.
Principally, the calibration procedure of Flying Triangulation (e.g. described in [1–4]) could be applied for the 3D movie camera as well. This was done at an early stage of the project. However, it soon turned out that the *accuracy* provided by this

F. Willomitzer, *Single-Shot 3D Sensing Close to Physical Limits and Information Limits*, Springer Theses,
https://doi.org/10.1007/978-3-030-10904-2_8

procedure (especially the *accuracy of back projections!*) is not sufficient if higher line numbers $N_L > 80$ are projected [5, 6]. Moreover, the procedure requires an expensive translation stage, is very time consuming, and prone to several alignment errors. The major problem is, that two *different* planar calibration targets are applied, which have to be changed during the calibration process. Nevertheless, the position, alignment and surface of both targets are considered as *equal*.

Due to these severe drawbacks, a new calibration procedure has been developed in [7–10]—initially for Flying Triangulation. The procedure significantly improves the calibration by the introduction of photogrammetric methods. The major improvements are:

- *Higher accuracy*, sufficient for the correct back projection of $N_L = 160$ lines.
- Usage of *only one calibration target* (planar screen), which also represents the *only* additional required hardware component.
- Robustness against misalignment of the calibration target.
- Robustness against deviations of the calibration target from a perfect plane.

In this thesis, the calibration procedure of [7–10] was adopted for the single-shot 3D movie camera. This required several (mainly algorithmic) modifications, e.g. the correct identification of up to 160 lines in one image or the implementation of a second camera. These modifications are described more detailed in the following. For a detailed description of the remaining basic steps, see [9, 10].

As an illustrating example, the calibration of the 3D movie camera prototype is shown, projecting a crossed line pattern with 160 vertical and 100 horizontal lines. Note: For measurements comprising a crossed pattern, vertical and horizontal lines are calibrated separately.

8.1.1 Image Acquisition

At the beginning of the calibration process, several images have to be acquired. The images are necessary for the *model-based internal calibration* of both cameras and the *model-free external calibration* between cameras and projector. As mentioned, only the calibration plate and the sensor itself is required for this process. The calibration plate consists of a white screen with an imprinted periodic pattern of small black dots.[1] During the image acquisition, the room should be as dark as possible. Light is only provided by the projector.

[1]For the prototype setup the utilized calibration plate measures about 360 × 260 mm. The plate material is 'Aluminum Dibond' laminated with an adhesive white paper. A pattern of 34 × 24 black dots with a diameter of about 1.5 and 10 mm distance between two dots is imprinted.

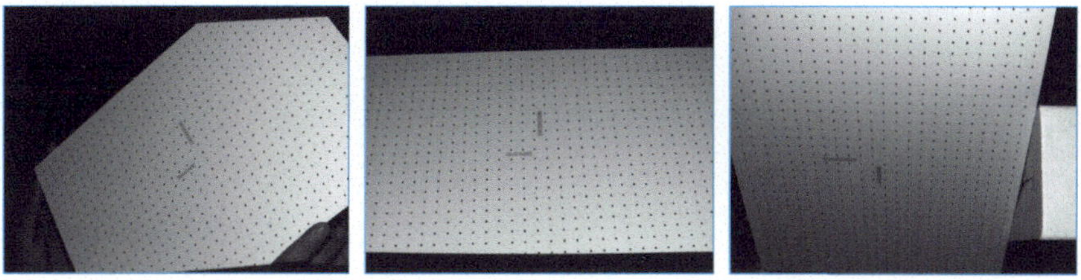

Fig. 8.1 Images for the internal calibration. Three examples, acquired with C_1

For the *internal calibration of both cameras*, the plate is placed under *different tilt angles* and *different rotations* at *different positions* in the measurement volume. The projector illuminates the plate with a homogeneous 'white picture'. Both cameras acquire an image of the illuminated plate. Figure 8.1 shows three example images acquired with C_1. 10 to 15 acquired images per camera are sufficient for good results, if tilt and rotation is varied properly.

Subsequently, the images for the *external calibration between cameras and projector* are acquired: The calibration plate is aligned (by hand) roughly perpendicular to the optical axis of the projector and is placed (also by hand) at J different depths within the measurement range. In the case of the 3D movie camera prototype, reasonable depth steps are about 10 mm, which results in $J = 11$ images for the measurement range of 100 mm. At *each* depth, the following pictures[2] are projected and an image is acquired with *each* camera:

- Homogeneous white picture
- Picture of 160 vertical lines
- Picture of 100 horizontal lines
- Picture of 1 vertical line
- Picture of 1 horizontal line

Of course, the *position of all lines* in the different *projector pictures* is *fixed* and does not change with depth. Only the line positions in *the camera images* change due to the perspective. The projected *single line* has to display the *same position* as one certain line in the 'multi-line' picture, e.g. *line 80*, resp. *line 50*. In Sect. 8.1.4 the 'single-line image' is used for index assignment.

The different depth positions of the calibration plate are illustrated in Fig. 8.2a. Figure 8.2b–f shows two example images for each projected picture. The two images are acquired at the depth $z = 470$ mm and $z = 530$ mm with camera C_1. As seen in Fig. 8.2e, f, the vertical lines virtually display no shift in the camera images of C_1 if

[2]The described projection of different pictures is only possible with a switchable projector. The utilization of a slide-projector (as recommended several times) would require modifications: Vertical and horizontal lines would be calibrated together as crossed pattern. For the 'white picture' a second light source (e.g. room light) has to be utilized. The single lines can be projected by a second (precisely adjusted) projector. Alternatively, one vertical and horizontal line in the crossed pattern is specially labeled.

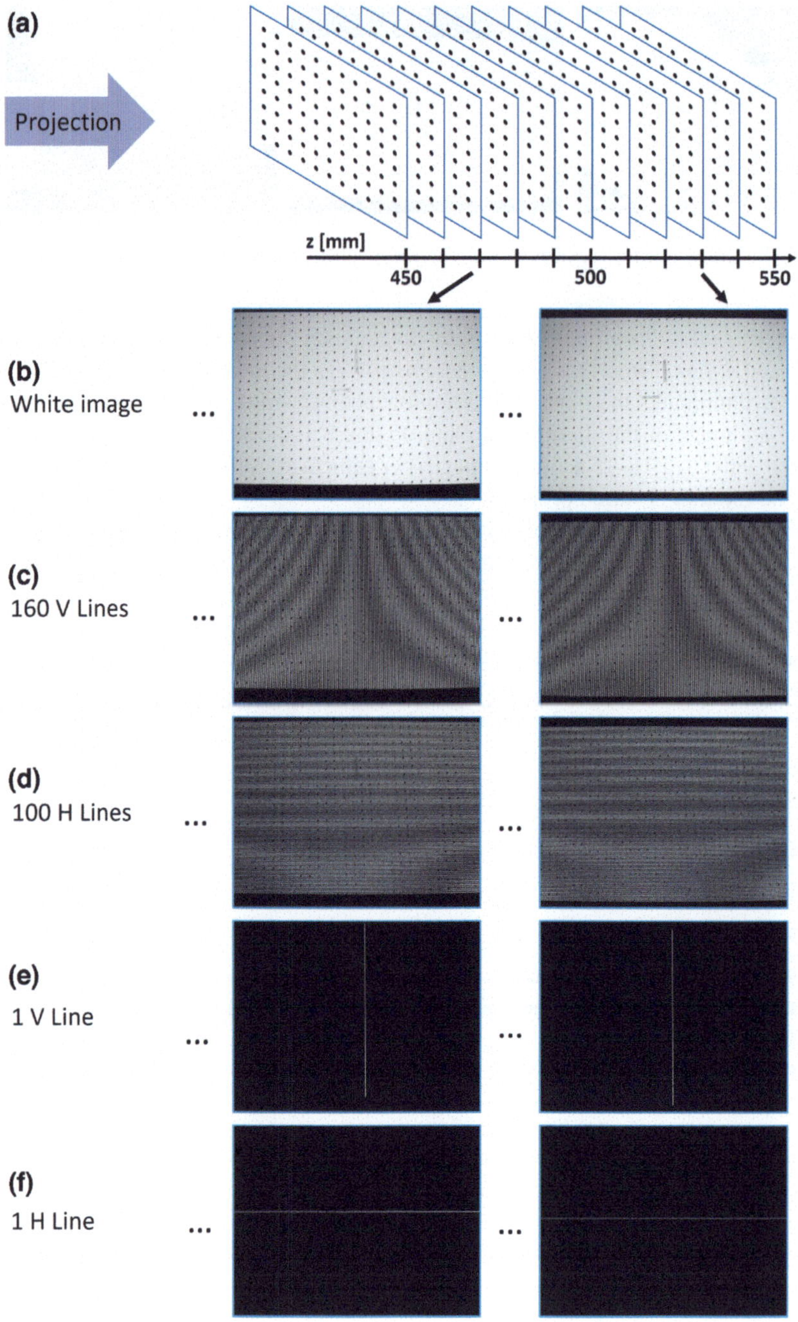

Fig. 8.2 Images for the external calibration. **a** Approximate depth positions of the calibration plate. **b–f** Examples for acquired camera images. For each projected picture, two images, taken at $z = 470$ mm and $z = 530$ mm with camera C_1, are shown

the depth is changed (small $\theta_{1,v}$), whereas the horizontal lines display a larger shift (large $\theta_{1,h}$). The images of camera C_2 (not shown here) display contrary behavior.

The next section explains the internal calibration of both cameras with photogrammetric methods as well as the resulting effects.

8.1.2 Internal Calibration

The (separate) internal calibration of both cameras yields the following major results:

- Exact location of all rays of sight in space.
- Exact (three dimensional) position in space of all calibration plate markers.

The calibration is performed with the photogrammetry software *Australis* [11]. The software uses the *expanded pinhole camera model* of Fraser [12, 13]. The camera images for the internal calibration are loaded into the software and the position of each marker (including its number) is determined for each image. Eventually, the output is calculated via optimization of the *RMS error* (root mean square) with *bundle adjustment* [14]. A detailed description of the procedure is given in [9]. According to the applied model of Fraser, the output for each camera consists of the following parameters:

- Interior camera orientation: c, x_p, y_p
- Radial lens distortion: k_1, k_2, k_3
- Decentring distortion: p_1, p_2
- In-plane correction: b_1, b_2

Due to the overdetermined set of equations, the bundle adjustment simultaneously allows a correction of the marker positions on the calibration plate. The markers may not be printed exactly at the correct positions on the plate and the surface of the plate may deviate from the perfect plane. The second part of the output is a *.txt*-file with the corrected three dimensional positions of the markers on the plate. In Sect. 8.1.5 these positions are used to approximate the true shape of the calibration plate.

The following sections explain the basic steps of the *external calibration*.

8.1.3 Resection

After the optimization procedure of the bundle adjustment, the rays of sight in space are assumed to be known for each camera. This allows for a localization in space of each object with *known geometry* with only one acquired image. In this case *'known geometry'* means *known distances of surface features*, e.g. the distances of markers on the calibration plate. This localization procedure is called *resection*.

In the second part of the image acquisition, images of the calibration plate with different illumination were acquired at $J = 11$ positions within the measurement volume. These different positions have to be known to enable the external calibration. The positions are determined via the resection of all acquired 'white images' (see Fig. 8.2b). The *.txt*-file with the corrected marker positions from above is used as distance-reference. Output of the resection is a *.txt*-file that contains the exact position (including tilt and rotation) of the calibration plate in space for each of the $J = 11$ different depths.

Important Note: This step is *one of the major improvements* of the new calibration method. In Flying Triangulation, the different depths of the plate were adjusted with a translation stage [2]. An unintentional tilt or deformation of the plate was not taken into account. Instead, the z-coordinate of the *entire plate surface* was considered as the z-value of the translation stage, e.g. $z = 500$ mm.

8.1.4 Line Maxima Evaluation and Indexing

In the next step, the intensity maximum of *each* line in *each* of the $J = 11$ acquired images of Fig. 8.2c, d is evaluated. The procedure (as well as the following steps) is only demonstrated for images with *vertical lines*. The evaluation of horizontal lines is analog.

As before, the intensity maximum is determined by Gauß approximation. Figure 8.3 summarizes the basic steps. For better visualization, a small image segment of 100×20 *pix*, located at the *left rim* of an image with multiple vertical lines (see Fig. 8.2c) is shown.

Figure 8.3a displays the raw image. Figure 8.3b shows the positions of the black markers, which are obtained from the 'white images' of Fig. 8.2b. In Fig. 8.3c, the intensity values are interpolated across the regions of the black markers. Note that this image is not used for the calculation of the Gauß profiles. It is only required

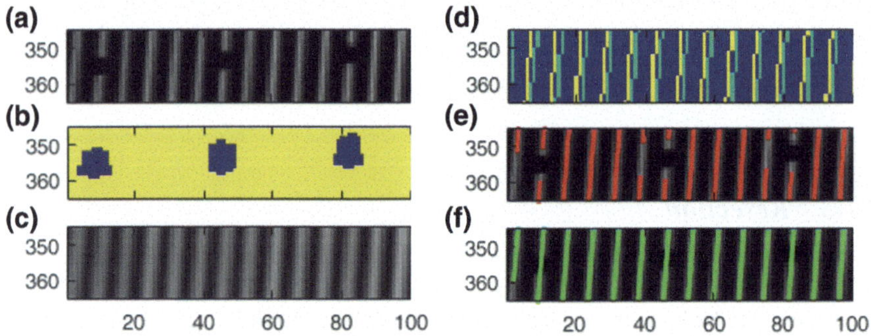

Fig. 8.3 Evaluation of line maxima in the calibration images

for the proper calculation of the boundary pixels for the Gauß fit (Fig. 8.3d) via differentiation. The green and yellow pixels in Fig. 8.3d represent the left and right side of a line profile. The Gauß profile of a line is only calculated if *both sides are present* and if *no marker is close-by*. Otherwise, the line profile may be cut obliquely, which would deliver false results. Figure 8.3e displays the calculated Gauß maxima in red.

As shown in Sect. 6.2, the calculated maxima display *statistical noise* $\delta x'$. For this reason, a polynomial of 5th degree is fitted trough all maxima of each line. For further processing, the values of this polynomial (and not the noisy maxima) are considered. In this case a polynomial fitting is permissible for two reasons: First, noise is a statistical process, which should not be calibrated into a sensor setup. Second, the lines on the planar surface of the calibration plate are a very smooth and can be approximated very precise with a proper polynomial. The polynomial fitting has the additional advantage that proper maximum values can also be generated at the positions of the markers by interpolation. Figure 8.2f shows the fitted line maxima in green.

Each of the evaluated maxima belongs to a certain *line index* which has to be determined in the next step. An index assignment by simply counting the lines from left to right (like in the Flying Triangulation calibration) is not possible. Lines may leave and enter the field of view, if the calibration plate is shifted. The method proposed in [6] is not practicable for higher line numbers $N_{L,v} \gtrsim 80$.

The problem is solved by using the 'single-line images' of Fig. 8.2e, f. Figure 8.4 illustrates the procedure: The position of the single line in the 'single-line image' is detected by simple maximum evaluation. The corresponding line in the 'multi-line image' with the *same position* is then assigned to the fixed index n.

Figure 8.4 shows the same image segment of single-line and multi-line images, acquired at the beginning ($z = 450$ mm) and the end ($z = 550$ mm) of the measurement range with camera C_2. The line shift between these images is more than *eight line distances*.

In the images of camera C_1, the line shift over the whole measurement range is *by definition less than one line distance*. These images (as well as the images with

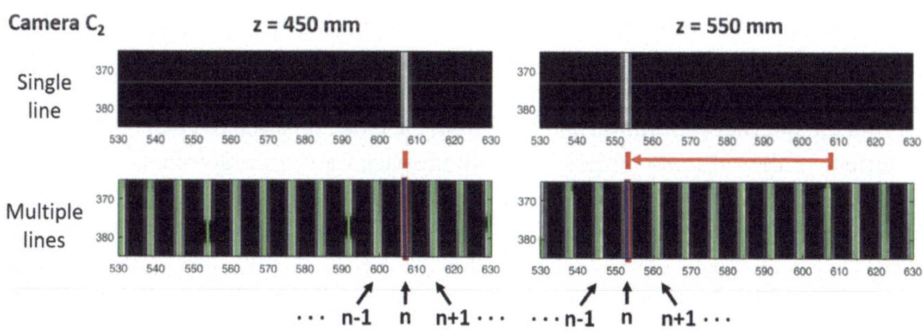

Fig. 8.4 Index assignment with the 'single-line image'

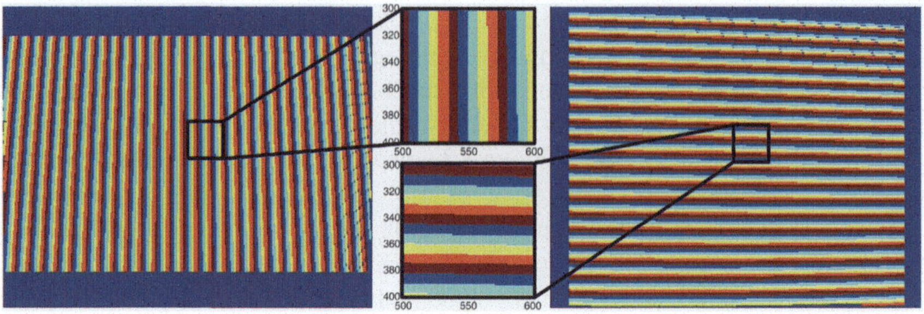

Fig. 8.5 Regions of uniqueness for the single-shot 3D movie camera prototype. Vertical regions on the chip of C_1, horizontal regions on the chip of C_2. For 160 vertical and 100 horizontal projected lines each region is about 7 pix wide

horizontal lines from C_2) are subsequently used to define[3] the 'regions of uniqueness' necessary for the unique evaluation of the first 'noisy' 3D model (see Sects. 4.1 and 8.2.2). Figure 8.5 shows the regions of uniqueness for both line directions.

8.1.5 Longitudinal Calibration

During the longitudinal calibration, a transformation $\zeta_{ny'}$ that transforms each lateral sub pixel precise line maximum value x' on the chip into a depth value z in space is calculated. Since multiple lines are projected and the lines are distorted, the transformation $\zeta_{ny'}$ has to be calculated separately for each line index n and each pixel-row y', whereas $1 \leq n \leq N_L$ and $1 \leq y' \leq N_y$ being integers.

It has been shown that a polynomial of 3rd degree is appropriate to describe the transformation [1–4]:

$$z = \zeta_{ny'}(x') = a_{ny'} + b_{ny'}x' + c_{ny'}x'^2 + d_{ny'}x'^3 \tag{8.1}$$

The coefficients are determined as follows: During the evaluation of the line maxima in the last section, a lateral maximum position $x'_{jny'}$ (float) was evaluated for *each index n* and *each row y'* in *each* of the *J acquired images*,[4] where $1 \leq j \leq J$ represents the integer image number.

From the internal calibration and resection, the exact position of each marker onto the surface of the calibration plate is known in space for each acquired image. These positions are now used to approximate the (not planar!) surface of the calibration plate at each of the J depth positions. The approximation with 'radial basis functions' (RBFs) is explained in [9]. Eventually, the rays of sight of the evaluated maxima

[3]Tilt of the calibration plate is compensated.

[4]Discussed example of the prototype setup with vertical lines: $N_L = 160$; $N_y = 682$; $J = 11$.

values are intersected with the approximated plate surface for each image j. This yields a value $z_{jny'}$, that uniquely corresponds to each $x'_{jny'}$.

The coefficients of $\zeta_{ny'}$ are finally evaluated by solving the following equation system separately for each index n and each row y' with the method of 'least squares':

$$\begin{pmatrix} 1 & x'_{1ny'} & (x'_{1ny'})^2 & (x'_{1ny'})^3 \\ 1 & x'_{2ny'} & (x'_{2ny'})^2 & (x'_{2ny'})^3 \\ \vdots & \vdots & \vdots & \vdots \\ 1 & x'_{Jny'} & (x'_{Jny'})^2 & (x'_{Jny'})^3 \end{pmatrix} \begin{pmatrix} a_{ny'} \\ b_{ny'} \\ c_{ny'} \\ d_{ny'} \end{pmatrix} = \begin{pmatrix} z_{1ny'} \\ z_{2ny'} \\ \vdots \\ z_{Jny'} \end{pmatrix} \tag{8.2}$$

8.1.6 Finalization

A *lateral calibration* (like in Flying Triangulation) is *no longer necessary* [9]. The lateral 3D coordinates x and y are calculated as follows: From each maximum position x' the corresponding z-value is calculated via Eq. (8.1). Eventually, the ray of sight of the image coordinate (x', y') is evaluated at the calculated z-value, which delivers the coordinate (x, y) in space.

The calculation of the rays of sight with the extended pinhole model implies distortion and orientation corrected metric image coordinates (x'_{corr}, y'_{corr}). These coordinates are simply projected through the optical center (distance c) to obtain the lateral coordinates in space:

$$\begin{pmatrix} x \\ y \end{pmatrix} = \frac{z}{c} \begin{pmatrix} x'_{corr} \\ y'_{corr} \end{pmatrix} \tag{8.3}$$

According to [9], x'_{corr} and y'_{corr} are calculated with the internal camera parameters determined in Sect. 8.1.2 via

$$\begin{aligned} x'_{corr} &= \widetilde{x'} + \frac{\widetilde{x'} \cdot dr}{r} + p_1 \cdot (r^2 + 2 \cdot \widetilde{x'}^2) + 2 \cdot p_2 \cdot \widetilde{x'} \cdot \widetilde{y'} + b_1 \cdot \widetilde{x'} + b_2 \cdot \widetilde{y'} \ , \\ y'_{corr} &= \widetilde{y'} + \frac{\widetilde{y'} \cdot dr}{r} + p_2 \cdot (r^2 + 2 \cdot \widetilde{y'}^2) + 2 \cdot p_1 \cdot \widetilde{x'} \cdot \widetilde{y'} \ , \end{aligned} \tag{8.4}$$

$\widetilde{x'}$, $\widetilde{y'}$, r and dr being given by:

$$\begin{aligned} \widetilde{x'} &= x' - x_p & r &= \sqrt{\widetilde{x'}^2 + \widetilde{y'}^2} \\ \widetilde{y'} &= y' - y_p & dr &= k_1 \cdot r^3 + k_2 \cdot r^5 + k_3 \cdot r^7 \end{aligned} \tag{8.5}$$

After the calibration, 3D points can be evaluated from each camera separately. In order to work together correctly, both cameras have finally to be transformed into a common coordinate system. By default, C_2 is transformed in the system of C_1. The related transformation is calculated from the resection data.

8.2 Evaluation

This section briefly lists and discusses the algorithmic steps for the evaluation of 3D data, using a concrete example acquired with the 3D movie camera prototype setup. The example images of C_1 and C_2 are two color images with crossed lines. The projected pattern consists of 142 vertical and 88 horizontal lines. The images are taken from an entire sequence. The corresponding 3D movie can be seen on [15, 16] (direct link: tinyurl.com/3DMovCam12).[5]

The considered image of C_2 is shown in Fig. 8.6. In the following only an image segment of 80×80 pix (see magnification window) is considered for better visualization of all steps. This segment is chosen intentionally to demonstrate the performance of the algorithmic steps for difficult conditions: A surface region with high inclination (nasal bone) is measured, which results in a clear deformation of the lines. According to the very low contrast, a proper evaluation of 3D points is not always possible.

As before, the following sections refer only to the evaluation of vertical lines. In order to evaluate horizontal lines, the image is turned by 90° and undergoes the same algorithmic procedure.

For the evaluation of 3D data, each image is converted into gray values. This step is not necessary if monochromatic cameras are used in the setup. The RGB values are required in Sect. 8.2.3 for the evaluation of the color texture.

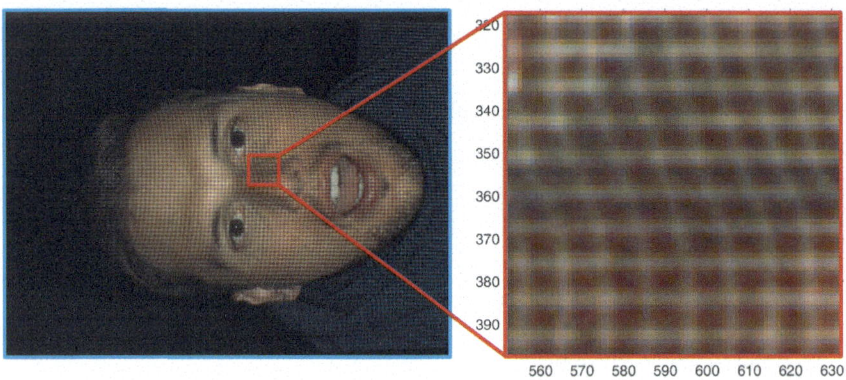

Fig. 8.6 Image frame used for the demonstration of the evaluation algorithm: Image of C_2 used for the final evaluation of the vertical line direction

[5]Alternative link: tinyurl.com/3DCam-012.

8.2.1 Separation of Line Directions

The method for the separation of both line directions was introduced in Sect. 7.1.1. Figure 8.7 summarizes the basic algorithmic steps for the vertical lines.

The line directions have to be separated in both acquired camera images. The lower part of Fig. 8.7 displays the considered region of 80×80 pix in the image of C_2, whereas the upper part of Fig. 8.7 shows the corresponding region for C_1. Figure 8.7b visualizes the 2D Fourier transform of the gray-converted raw image (full image) of Fig. 8.7a. The window to mask the Fourier spectrum (red) has to be chosen wider for the images of C_2, since the vertical lines display a larger deformation. 2D Fourier back transformation of the masked spectrum leads to Fig. 8.7c. These images are taken to calculate the mask for the vertical line direction via differentiation. Figure 8.7d visualizes the calculated mask, which is already applied on the raw images of Fig. 8.7a. These images are used in the following steps for the evaluation of the sub pixel precise Gauß maxima.

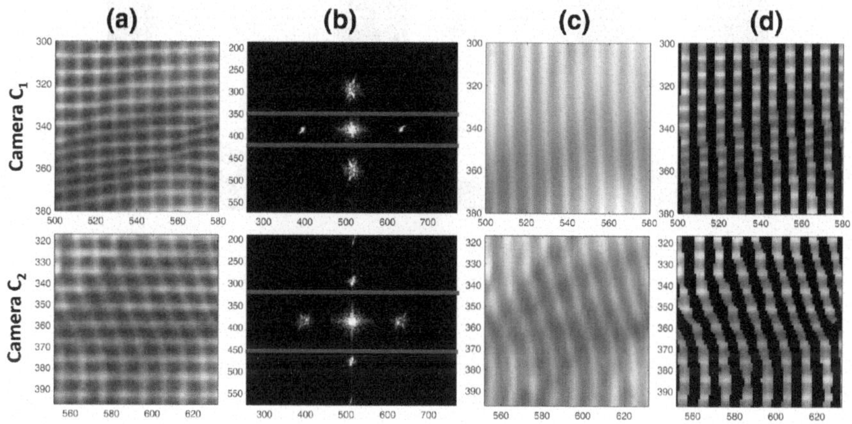

Fig. 8.7 Separation of line directions in the Fourier domain shown for the images of both cameras. **a** Original image. **b** 2D Fourier transform (magnified) and applied masking window. **c** 2D Fourier back transformation of the masked fourier spectrum. **d** Mask calculated from (**c**), applied on (**a**)

8.2.2 Indexing and Evaluation of the 'Coarse' 3D Model

The next step is the evaluation of a noisy 3D model with known indices from the image of C_1. First, the sub pixel precise Gauß maxima are evaluated for *each* pixel-row of *each* line. To approximate a line profile with the Gauß function of Eq. (6.1), the related algorithm requires the definition of a left and a right border. Within these borders, *one* full line profile (4–6 pix width) is supposed to be located. The lateral positions of the borders are obtained from the mask of Fig. 8.7d. Figure 8.8a visualizes

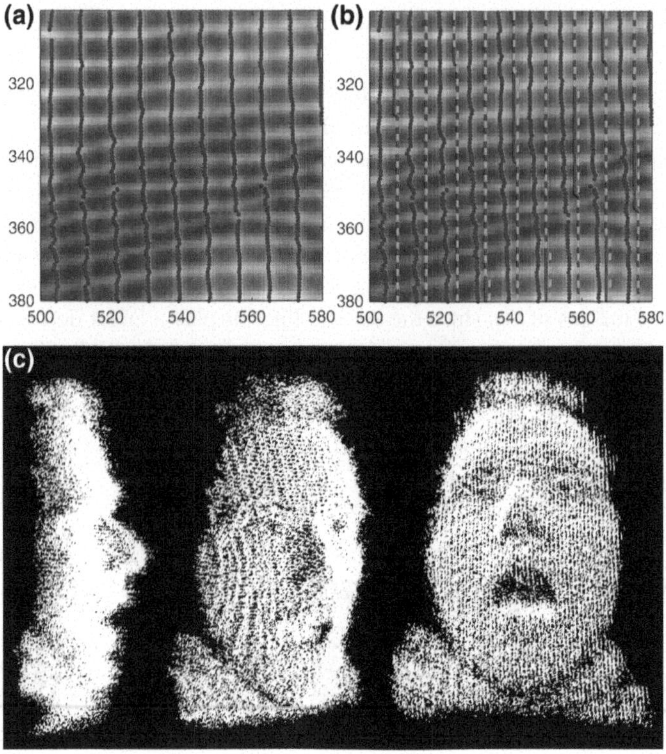

Fig. 8.8 Index assignment and evaluation for the coarse 3D model. **a** Evaluated Gauß maxima for the vertical lines. **b** Gauß maxima within the regions of uniqueness. **c** Coarse 3D model, shown from three different viewpoints

the evaluated Gauß maxima positions, plotted onto the raw image of C_1 in Fig. 8.7a. These positions are not assigned to a line index n so far. The assignment requires the *regions of uniqueness*, which were determined in Sect. 8.1.4. Figure 8.8b displays the right (red lines) and left (green dashed lines) borders of the regions of uniqueness for the considered image segment. Each region represents a different line index n. The index n is assigned to the corresponding sub pixel precise line maximum x' for each pixel-row y'. Eventually the space coordinates (x, y, z) are calculated with Eqs. (8.1) and (8.3). Figure 8.8c shows the evaluated 3D model from three different viewpoints. Due to the small triangulation angle θ_1, the noise on the 3D data is several millimeters (also see Fig. 8.9d). The shown model already contains the 'coarse lines' from the other line direction, i.e. the horizontal lines from the image of C_2.

Important Note: In Flying Triangulation, the Gauß maxima are not evaluated within the borders shown in Fig. 8.7d, but within the borders of the regions of uniqueness. For a large number of projected lines this results in a huge drawback: The *full line profile* (and not only its maximum) has to be permanently located within the region of uniqueness. This consumes additional space of $\sim 4\ pix$ for each region of uniqueness on the chip, which is lost for the measurement depth, resp. the number of projected lines, resp. precision (see Eq. (4.4)). If, as in the shown case, the width of a region of uniqueness is only 8–9 pix or even smaller (if 160 vertical lines are projected), 4 pix is a significant quantity. The need of *different-sized border-intervals* for Gauß evaluation and indexing was already identified in [5].

8.2.3 Index Back Projection and Evaluation of the Final 3D Model

The 'coarse' 3D model of Fig. 8.8c is back projected line-wise onto the chip of the other camera. In the case of the vertical lines, this is C_2. Mathematically, the back projection is performed by the inverse operation of Eq. (8.3). The longitudinal calibration is not required. Since the back projected 3D model is unique, each back projected line is assigned to a certain line index n.

Simultaneously, the sub pixel precise Gauß maxima are evaluated for C_2, also from the masked image of Fig. 8.7d. Figure 8.9a displays the evaluated Gauß maxima (blue) and the back projected points of the 'coarse' 3D model. Each color represents a different line index n. Eventually, a line index is assigned to each Gauß maximum. The back projection with the *smallest distance* to each maximum defines its index n. The maximal allowed distance (the 'search range') of a back projection is given by the 'scattering condition' of Eq. (6.12), i.e. it should be smaller than half the line distance. A typical search range for objects like human faces is about 2–3 pix to each side, which is not much wider than the intensity profile of the line. Figure 8.9b visualizes the indexed Gauß maxima in the image of C_2, with each color representing a different index n.

Due to the accumulation of different errors, it may happen that a few single back projections are landing close to a 'false' line. This would result in a false index assignment. Although this happens rarely, the error can be prevented by considering *small line segments* instead of single line points. The corresponding index is then assigned via a 'majority decision' for the whole segment. However, this is only possible if the object displays enough smoothness, since information from a small neighborhood is required.

The 3D data are calculated analog to the last section using the calibration parameters for C_2. Figure 8.9c displays the evaluation result shown from the same viewpoints as the 'coarse model' in Fig. 8.8c. The visualized 3D model already contains the points from the horizontal lines evaluated from the image of C_1. Figure 8.9d compares the final evaluation result (white) with the 'coarse model' after the first evaluation step (red).

In a last (optional) algorithmic step, the final 3D model is textured. The basic procedure was explained in Sect. 7.2.

In simplified terms, a three dimensional texture coordinate (r, g, b) is attached to each space coordinate (x, y, z). In the given example of vertical lines, the (r, g, b)

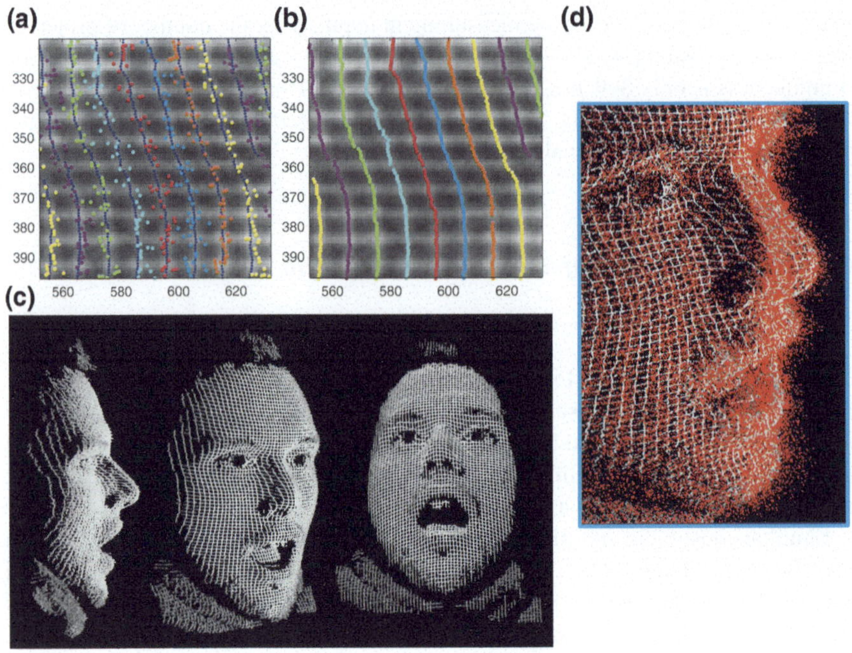

Fig. 8.9 Index assignment and evaluation for the final 3D model. **a** Evaluated Gauß maxima for the vertical lines (blue) and back projected points of the coarse 3D model. **b** Gauß maxima with assigned index. **c** Final 3D model shown from three different viewpoints. **d** Comparison of coarse (red) and final 3D model (white)

coordinate for each point is evaluated from the single color channels of the image of C_2. Figure 8.10a visualizes the pixels located at the positions of the the Gauß maxima, where the (r, g, b) value is measured. For monochromatic images, (r, g, b) has three equal entries. Figure 8.10b shows the final 3D model with different texturing: untextured, textured with gray values and textured with color values.

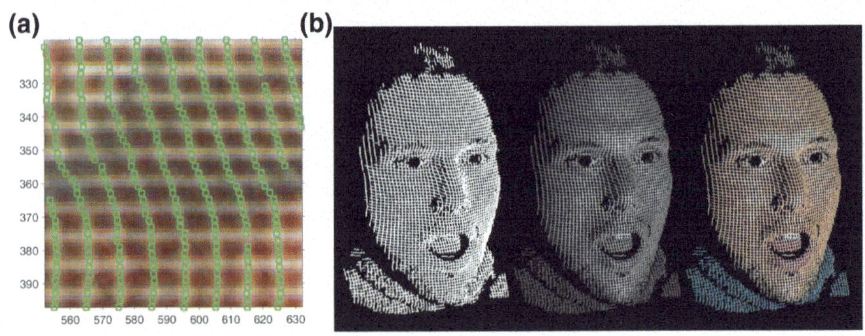

Fig. 8.10 Texturing of the final 3D model. **a** Pixels on the image of C_2, where the (r, g, b) coordinate is evaluated. **b** Final 3D model with different texturing

References

1. F. Willomitzer, FlyFace – 3D-Gesichtsvermessung mittels "Flying Triangulation". Diploma Thesis, University Erlangen-Nuremberg (2010)
2. O. Arold, Flying Triangulation - handgeführte optische 3D Messung in Echtzeit. Dissertation, University Erlangen-Nuremberg (2013)
3. P. Vogt, Beleuchtungsoptimierung für einen bewegungsunempfindlichen 3D-Sensor. Diploma Thesis, University Erlangen-Nuremberg (2008)
4. Z. Yang, Ein miniaturisierter bewegungsunempfindlicher 3D-Sensor. Master Thesis, University Erlangen-Nuremberg (2009)
5. F. Willomitzer, S. Ettl, C. Faber, G. Häusler, Single-shot three-dimensional sensing with improved data density. Appl. Opt. **54**(3), 408–417 (2015)
6. M. Kuch, Neue Indizierungsmethoden für die Multilinientriangulation. Master Thesis, University Erlangen-Nuremberg (2014)
7. M. Schröter, Kalibrierung des optischen 3D-Sensors "Flying Triangulation". Diploma Thesis, University Erlangen-Nuremberg (2012)
8. M. Schröter, F. Willomitzer, E. Olesch, O. Arold, S. Ettl, G. Häusler, Calibration of "Flying Triangulation". in *Proceedings of the 113th DGaO Conference* (2012), p. 25
9. F. Schiffers, Kalibrierung von Multilinientriangulationssensoren. Bachelor Thesis, University Erlangen-Nuremberg (2014)
10. F. Schiffers, F. Willomitzer, S. Ettl, Z. Yang, G. Häusler, Calibration of multi-line-light-sectioning. in *Proceedings of the 115th DGaO Conference* (2014), p. 12
11. Photometrix, Australis (2017). www.photometrix.com.au/australis/. Accessed 15 May 2017
12. C.S. Fraser, Digital camera self-calibration. ISPRS J. Photogramm. Remote. Sens. **52**(4), 149–159 (1997)
13. C.S. Fraser, Automatic camera calibration in close range photogrammetry. Photogramm. Eng. Remote. Sens. **79**(4), 381–388 (2013)
14. B. Triggs, P.F. McLauchlan, R.I. Hartley, A.W. Fitzgibbon, Bundle adjustment a modern synthesis. in *Vision Algorithms: Theory and Practice* (Springer, Berlin, 2000), pp. 298–372
15. Video file repository for Florian Willomitzer's Dissertation (urn:nbn:de:bvb:29-opus4-85442). University Erlangen-Nuremberg. nbn-resolving.de/urn:nbn:de:bvb:29-opus4-85442. Accessed 31 May 2017
16. Osmin3D, YouTube-channel of the OSMIN research group at the University Erlangen-Nuremberg. www.youtube.com/user/Osmin3D. Accessed 15 May 2017

Chapter 9
Results

Several examples for *single-shot 3D models*, acquired with the 3D movie camera prototype were already shown in this thesis, for the sensor configuration with unidirectional lines as well as for the configuration with crossed lines. This will not be repeated here.

This chapter is *focused on 3D movies*. It explains the acquisition process, shows examples and discusses proper viewing methods. As already noted, all shown 3D models display only raw data, directly delivered from the sensor. No post processing like interpolation or smoothing is applied. In the last section, however, further processing steps (e.g. necessary for 3D printing) are discussed.

9.1 Acquisition of a 3D Movie

A 'single-shot 3D movie' consists of a series of subsequently acquired independent single-shot 3D models. Figure 9.1 visualizes the acquisition procedure. For a 3D movie of a human face, the person to be measured is placed in front of the sensor. The projected pattern (in this case crossed pattern) is switched on and *remains on* for the whole acquisition process (no strobe illumination). Each of the synchronized cameras acquires a series of images from which the 3D models are evaluated. Currently, the evaluation procedure is still performed offline. A real time evaluation with interactive feedback on the screen is work in progress.

F. Willomitzer, *Single-Shot 3D Sensing Close to Physical Limits and Information Limits*, Springer Theses,
https://doi.org/10.1007/978-3-030-10904-2_9

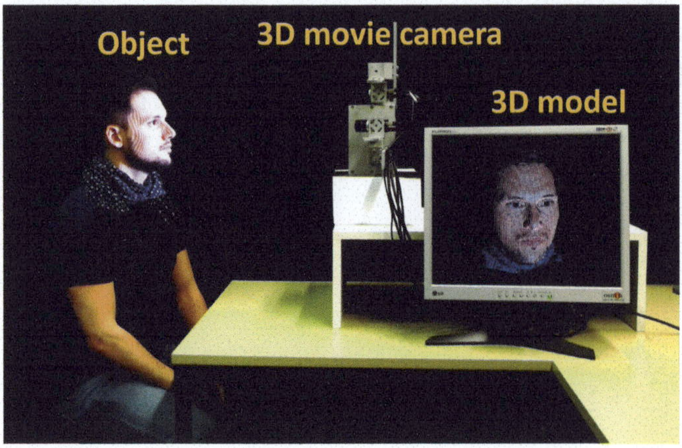

Fig. 9.1 Acquisition of a 3D movie with the single-shot 3D movie camera

9.2 How to View 3D Movies

While playing a 3D movie, the moving 3D model to be displayed on the monitor can be freely turned, tilted or magnified. A view from each perspective is possible. Hence, 3D movies are also often considered as 'free-viewpoint' movies. Figure 9.2 shows a person watching a 3D movie and changing the viewpoint with the mouse. The software for the visualization was written within the framework of this project. The figure displays a single frame of a video that can be watched in its full length on [1, 2] (direct link: tinyurl.com/3DMovCamView).[1] As an alternative to mouse navigation, the face of the observer could be tracked by a webcam and the viewing perspective could be changed according to the position of the observer. This gives the impression to freely 'look around' the 3D model.

[1] Alternative link: tinyurl.com/3DCam-view1.

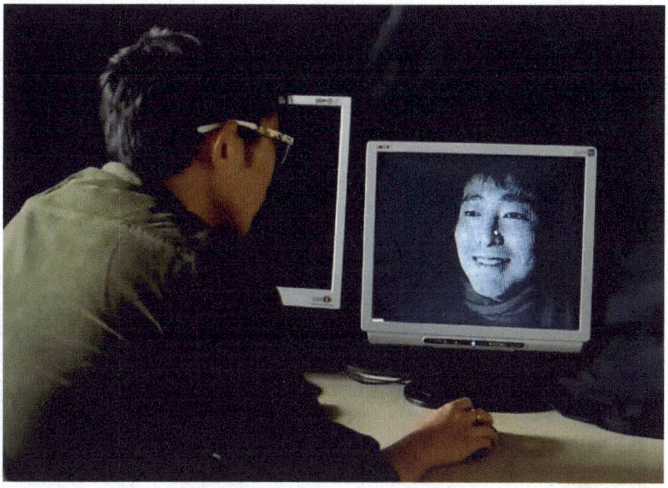

Fig. 9.2 Examination of a 3D movie. The viewpoint can be chosen freely with the mouse while the movie is playing

It is emphasized again, that such *free-viewpoint 3D movies* are not related to cinema '3D movies' (which should be better considered as *'2.5D movies'*). In these movies the 3D impression is only simulated and the viewpoint cannot be changed. Real 3D data are not present.

9.3 Examples for 3D Movies

A printed thesis cannot depict 3D movies in their entirety. The subsequent figures visualize a few acquired 3D movies the following way: From the whole movie sequence, three single 3D frames are picked, which represent three different points in time of the 3D movie. Eventually, each 3D frame is shown from three different viewpoints.

Additionally, a continuative hyperlink is given to each figure. The link leads to a video that shows the moving 3D model from predefined viewpoints.

3D Movie showing a Human Face with Gray-Texture (Crossed Lines)
Full video on [1, 2] (direct links: tinyurl.com/3DMovCam01 or tinyurl.com/3DCam-001).

Fig. 9.3 3D movie showing a human face with gray-texture

3D Movie showing a Human Face with Color-Texture (Crossed Lines)
Full video on [1, 2] (direct links: tinyurl.com/3DMovCam11 or tinyurl.com/3DCam-011).

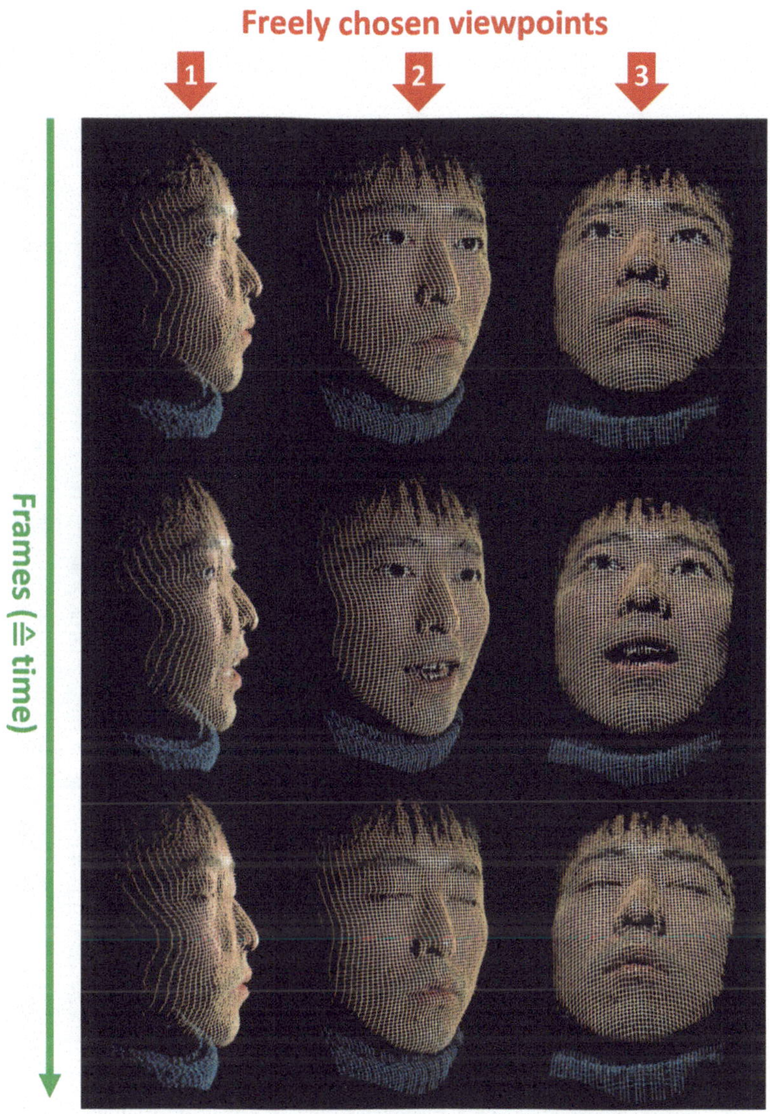

Fig. 9.4 3D movie showing a human face with color-texture

3D Movie showing a Jumping Ball with Gray-Texture (Crossed Lines)
Full video on [1, 2] (direct links: tinyurl.com/3DMovCam09 or tinyurl.com/3DCam-009).

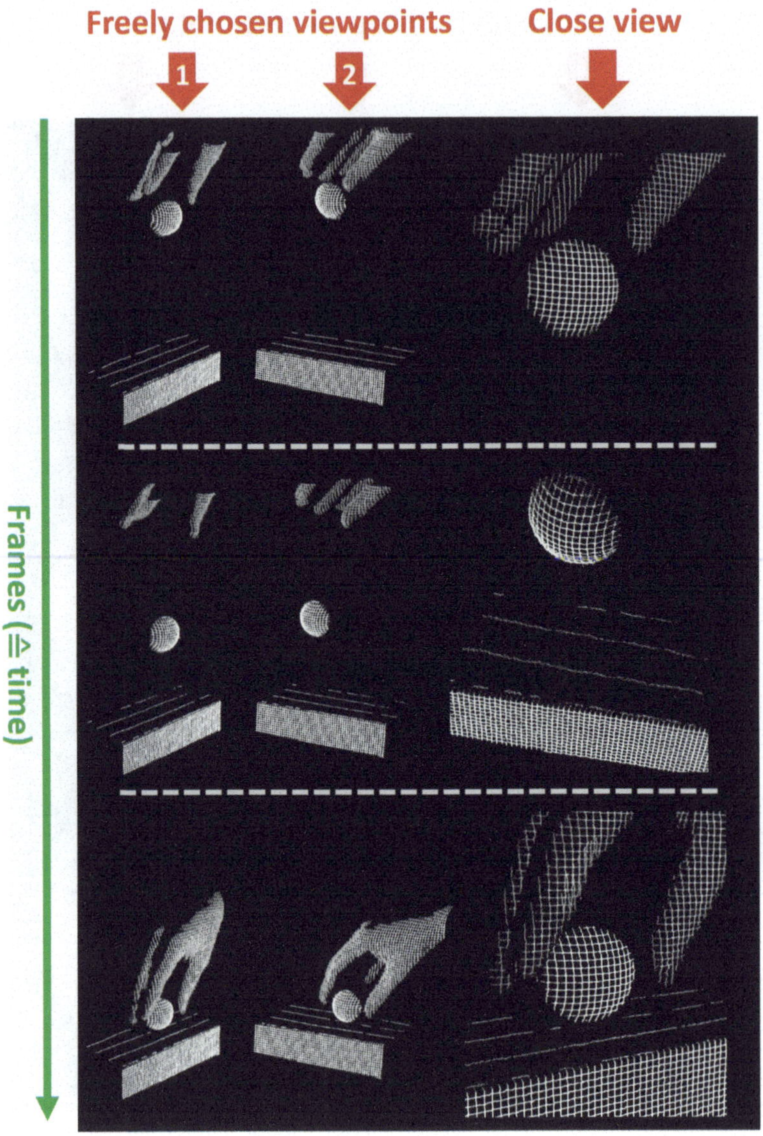

Fig. 9.5 3D movie showing a jumping ball with gray-texture

More Examples for 3D Movies

All videos can be seen on [1, 2].

Table 9.1 More examples for 3D Movies

Thumbnail	Object	Lines	Texture	Direct links
	Human face	Unidirectional	Gray	tinyurl.com/3DMovCam08 or tinyurl.com/3DCam-008
	Folded paper	Unidirectional	Gray	tinyurl.com/3DMovCam07 or tinyurl.com/3DCam-007
	Human face	Crossed	Gray	tinyurl.com/3DMovCam10 or tinyurl.com/3DCam-010
	Human face	Crossed	Gray	tinyurl.com/3DMovCam06 or tinyurl.com/3DCam-006
	Human face	Crossed	Color	tinyurl.com/3DMovCam12 or tinyurl.com/3DCam-012

9.4 Further Optional Processing Steps

This thesis is not about post processing of 3D point clouds. As shown, the single-shot 3D movie camera provides raw data point clouds with sufficient density and precision to be played as a movie. For this reason, post processing is generally nonessential.

In certain situations, however, post processing of 3D point clouds is unavoidable. Perhaps the best-known example is *3D printing*, which requires a so-called *'watertight' 3D model*, consisting of *a filled triangular mesh*.

The dense data structure of the acquired raw 3D models enables an easy execution of the necessary steps, which are visualized in Fig. 9.6: The acquired raw point cloud (Fig. 9.6a) is interpolated and thinned out. The related software was also written within the framework of this project, but perhaps other software (e.g. open source) is available as well. The resulting 3D model is shown in Fig. 9.6b. After the evaluation

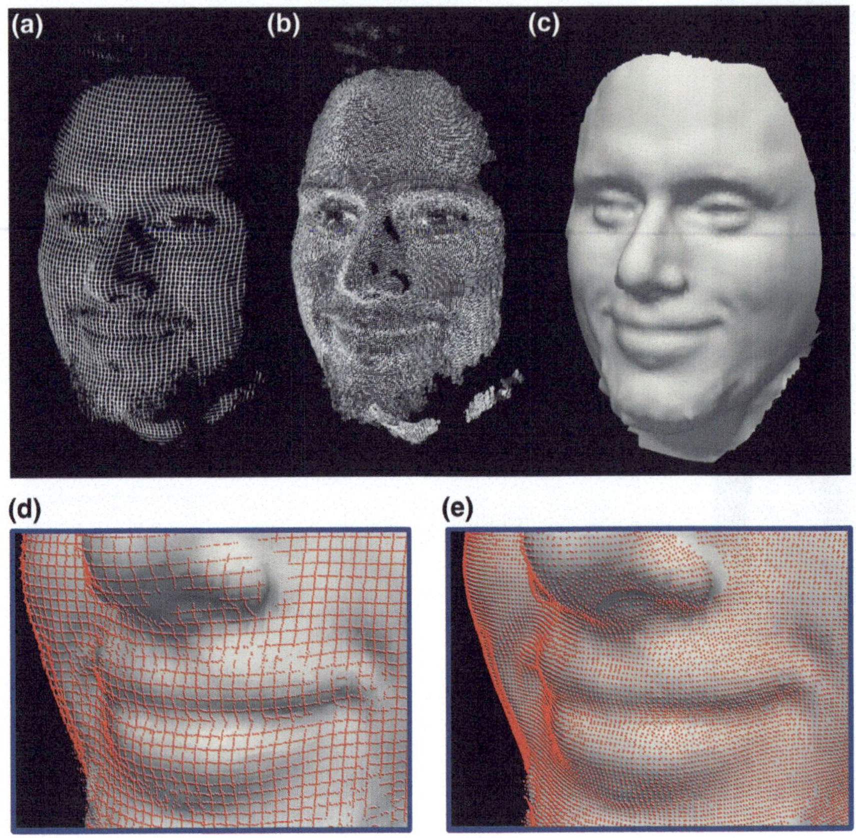

Fig. 9.6 Generation of a triangular mesh from the raw point cloud. **a** Raw point cloud delivered by the 3D camera. **b** Interpolated and thinned point cloud. **c** Filled triangular mesh. **d** and **e** Point clouds of (**a**) and (**b**) mapped onto the filled triangular mesh

of the surface normals, the triangular mesh is generated. For the shown example, this was done with the free software *'CloudCompare'* [3]. The resulting 'watertight' mesh is shown in Fig. 9.6c. Figure 9.6d, e compare the raw 3D point cloud delivered by the 3D camera, and the interpolated, thinned point cloud. Both point clouds are mapped onto the 'watertight' triangular mesh.

References

1. Video file repository for Florian Willomitzer's Dissertation (urn:nbn:de:bvb:29-opus4-85442). University Erlangen-Nuremberg. nbn-resolving.de/urn:nbn:de:bvb:29-opus4-85442. Accessed 31 May 2017
2. Osmin3D, YouTube-channel of the OSMIN research group at the University Erlangen-Nuremberg. www.youtube.com/user/Osmin3D. Accessed 15 May 2017
3. CloudCompare, [GPL software] (2017). www.cloudcompare.org. Accessed 15 May 2017

Chapter 10
Comments, Future Prospects and Collection of Ideas

This chapter summarizes complementary remarks and collects several ideas for a possible improvement of the single-shot 3D movie camera.

Section 10.1 is an addendum to the discussion of the state of the art in Chap. 3. It deals with sensor principles using an approach related to the approach developed in this thesis. Section 10.2 deals with a current problem of the 3D movie camera and introduces options for a solution. Currently, these options are only first ideas. An evaluation and implementation of these ideas is planned for the future. Section 10.3 introduces an additional idea for the improvement of the principle, namely the further increase of the projected pattern frequency. Section 10.4 discusses the question whether 3D movies of very fast processes can be acquired even with relatively 'slow' cameras.

10.1 State of the Art II: Sensors Using a Related Solution Approach

The *correspondence problem* is one of the *key problems* to be solved for *each* triangulation principle that aims for a high data density. In general, the solution of the correspondence problem requires the additional exploitation of modalities, such as *space*, *time*, *color* or *polarization*. Several examples therefor are discussed in Chap. 3.

In this thesis a novel solution approach was presented that exploits another modality: *Perspective*. The related information is provided by additional cameras in the setup. As the reader might guess, this source of additional information is not exclusively used by the single-shot 3D movie camera. Several authors recognized the benefits of applying more than one camera in their setups to exploit perspective information.

This section is complementary to the discussion about state-of-the-art triangulation sensors in Chap. 3. It displays how perspective information is used in other

F. Willomitzer, *Single-Shot 3D Sensing Close to Physical Limits and Information Limits*, Springer Theses,
https://doi.org/10.1007/978-3-030-10904-2_10

triangulation methods to solve (or extenuate) the correspondence problem with the help of a few examples. Most of the approaches mentioned were actually developed simultaneously to the 3D movie camera—a situation often occurring when demand is on the rise. To understand the similarities and differences, the related approaches are intentionally discussed here *after* the introduction of the concept used in the 3D movie camera.

In PMT (Phase Measuring Triangulation) approaches, perspective information is exploited more frequently. The main goal is the *reduction of the number of sequentially acquired images*. If the correspondence problem is solved by additional perspective information, no multi-frequency phase shifts or additional gray code sequences have to be acquired. In the best case, a unique 3D model can be generated from only 3 subsequent images, which significantly speeds up the measurement process. Related approaches can be found e.g. in [1–5]. The authors of [1–4] even use the combination of a small and a large triangulation angle. In [1], an evaluation method is proposed where the 3D points, generated from both cameras are compared in space. This method is very similar to the proposed 'alternative' for the index-back-projection-approach, which had been discussed in Chaps. 5 and 6.

However, although the sequence of required images can be shortened, related approaches are still *not single-shot*. In [3, 4], a 'temporal sliding window approach' is used to generate 3D models with the same rate as the camera image acquisition. Nevertheless, at least three sequential images are required for one 3D model. Fast moving objects will cause motion artifacts. In order to minimize these artifacts, elaborate hardware as fast switchable projectors and high speed cameras are required. In exchange, the acquired data density is virtually 100%.

Multi-camera approaches can also be used in Fourier transform profilometry. As discussed, illuminated object surfaces actually have to be 'smooth' in order separate the frequency bands. Theoretically a second camera mounted at a small triangulation angle θ_1 can ensure sufficient 'smoothness', as only minor pattern distortions are visible in the related image. Eventually, the directly evaluated phase could be matched (i.e. 'back projected') to the image of the second camera.

In [6] a two camera approach is proposed that follows the above idea in a certain way. However, since θ_1 is not defined properly, a unique matching of the fringe-phase is not possible. The authors match the phase with an iterative approach based on the coarse data of the disparity map. This requires the acquisition of a second pair of images with different illumination. In this case the single-shot ability is lost.

An example for a *real* single-shot approach based on additional perspective information, was already proposed about 10 years ago. The authors of [7] were probably the first to have brought up a very important question: *Can a temporal sequence of K images be replaced by K simultaneously acquired images from K cameras?* An answer is demonstrated in the paper: A pattern with binary stripes is projected onto the object surface and the K cameras are aligned so that the length of the triangulation base between projector and cameras increases with 2^K. The method strongly resembles the tomographic triangulation experiment with the difference to be considered that the cameras are aligned systematically.

For a unique decoding of M depth values, $\log_2(M)$ cameras are required. For a setup, comparable to the 3D movie camera with at least $\Delta Z/\delta z \geq 500$ depth values, the method of [7] would need more than 9 cameras.

10.2 Current Drawbacks and Possible Solutions

The concept of the single-shot 3D movie camera delivered a significant advantage over conventional multi-line triangulation setups: Due to the proper utilization of a second camera, the line number N_L, the unique measurement range ΔZ and the precision ($\sim \sin\theta$) are no longer coupled by Eq. (4.4). However, Eq. (4.4) still plays an important and limiting role for the sensor concept: The upper limit for the effective measurement range ΔZ_e is still defined via Eq. (4.4) by the line number and the small triangulation angle θ_1.

An example: For the prototype setup with $N_L = 160$ unidirectional lines and θ_1 approximately 1°, the resulting $\Delta Z_1 = \Delta Z_e$ is only about 100 mm (see Table 6.1). This depth is approximately equal to the depth of field of the projector, meaning a larger unique measurement range ΔZ_1 would not improve the effective measurement range ΔZ_e in this case. Moreover, $\Delta Z_1 = 100$ mm is sufficient for the intended measurement object—human faces.

However, a proper measurement requires a precise positioning of the object within ΔZ_1. This could become a problem if the sensor is used in everyday situations. The problem gets even worse if the sensor is guided by hand. Other measurement scenes might require even a much larger ΔZ_1 (projector with sufficient depth of field assumed).

Similar to Flying Triangulation, the current sensor produces outliers if the object surface is measured outside of ΔZ_1. However, since the 3D movie camera displays a much higher data density in each single-shot, outliers can be distinguished much easier from the correct 3D model. Computer vision methods provide several post processing approaches for an automatic detection. These approaches are not to be discussed here any further.

Instead, this section introduces *conceptual ideas* of how outliers in the 3D raw image can be *avoided* by further increasing the effective measurement range. One idea related to this question was already introduced in [8]. However, this idea seems practical for only a relatively limited number of $N_L \lesssim 80$ lines and is not further considered.

For the following considerations, it is assumed that 3D data should be acquired with the same precision as in the chapters before, meaning the largest triangulation angle in the setup must still display $\theta_2 = 9°$.

10.2.1 'Sensor-Cascading'

This idea is illustrated for the sensor configuration with unidirectional lines, but can be applied for crossed lines as well.

In Chap. 6 it was discussed that both triangulation angles cannot be chosen independently from each other. The fixed value of $\theta_2 = 9°$ puts a lower limit on θ_1 respectively an upper limit on ΔZ_1. These limits are caused by the lever between both sub-sensors (sensor 1 and sensor 2). For a very small θ_1, the lever would be too high to fulfill all required conditions (e.g. the 'scattering condition' of Eq. (6.12)).

The 'sensor-cascading' idea intents to *extenuate the lever between two sensors* by the introduction of a third sensor with an *intermediate triangulation angle* θ_3, with the result of $\theta_1 < \theta_3 < \theta_2$. The index back projection is first performed between sensor 1 and sensor 3. Eventually, the 3D data from sensor 3, which display lower noise, are back projected onto the chip of sensor 2, where the final dataset is evaluated. The unique measurement range of the whole sensor system is still determined by θ_1, which can now be *chosen smaller* due to the reduced lever.

Of course, the concept of 'sensor-cascading' allows to add even more sensors with additional triangulation angles $\theta_4, \theta_5, \theta_6, ...$ Moreover, the cascading also works for the *direct comparison of points in 3D space*, which was considered as an alternative to the index-back-projection-approach.

10.2.2 Extended 'Nonius-Method'

For the sensor configuration with unidirectional lines, the following idea also requires the introduction of a third camera enclosing an angle θ_3 with the projection. θ_3 can be larger or smaller as θ_1, but commonly should not be larger as θ_2. *Neither sensor 1 nor sensor 3 have to display unique data!* The correct (noisy) dataset together with the correct index information is evaluated with sensor 1 and sensor 3 via the *'nonius-method'*, which has been discussed in Sect. 6.4.2. This dataset is eventually back projected onto the chip of C_2 for the evaluation of the final (precise) dataset. For a proper selection of θ_1 and θ_3, an effective measurement range ΔZ_e much larger than ΔZ_1 or ΔZ_3 can be achieved. Compared to the 'nonius-method' for unidirectional lines with only two cameras, this method offers the advantage that the two cameras, used for the determination of the measurement depth (C_1 and C_3) are not used for the evaluation of the final 3D dataset. Hence, the relation of θ_1 and θ_3 can be optimized to reach the maximal effective measurement range. If this optimization is done by the 'large-small' configuration (see Sect. 6.4.2), the method becomes very similar to the 'sensor-cascading' approach of the previous section.

For the sensor configuration with crossed lines, the 'nonius-method' can be performed in another interesting way *not requiring a third camera at all.* Since each line direction should display the same precision in the final dataset, the 'large' triangulation angles for both directions must be fixed, i.e. $\theta_{2,v} = \theta_{1,h} = 9°$. The ratio of the remaining angles in the setup ($\theta_{1,v}$ and $\theta_{2,h}$) is selected in the same manner as discussed above, namely to generate a large effective measurement range. Eventually, a set of unique 3D points is evaluated from the related sensors *at the crossing points of the pattern*. The evaluated index information is transferred from each crossing point to the adjacent line segments. Note: This operation requires spatial neighborhood. Line segments which are not connected to crossing points cannot be indexed correctly. However, such segments should be very rare according to the large number of projected lines (i.e. density of crossing points). Of course, this drawback can be avoided by the introduction of additional cameras in the setup. Finally, the generated unique but noisy 3D model is back projected onto the chips of both cameras, where the final precise 3D model is evaluated by triangulation with $\theta_{2,v}$ and $\theta_{1,h}$.

10.2.3 Other Possibilities

There are several other possibilities for enlarging the effective measurement range. Intuitive approaches already having been discussed in this thesis are:

- *Variation of the pattern period.* This method can be very effective to increase the effective measurement range, especially if a crossed line pattern is projected. As discussed in Sect. 3.3.5, a similar approach is followed e.g. by the authors of [9].
- *Spatial codification of (several) lines.*
- *Color codification of (several) lines.*

The drawbacks of these methods have been elaborated already.

Note that the introduced ideas of the previous sections are not limited to sensor configurations with a unidirectional or a crossed pattern. Section 7.1.4 revealed that a sensor setup with K cameras projecting lines in R directions comprises theoretically up to $K \times R$ triangulation sensors. The tasks of these sensors can be freely chosen. If a large effective measurement range is desired, a subset of the sensors can be used only to enlarge the effective measurement range. For example, lines with a spatial codification can be projected in a third or fourth pattern direction. These lines can solely be used for the enlargement of the measurement range and not for the generation of 3D data.

10.2.4 The Role of Epipolar Geometry

The reader might ask why the epipolar geometry is not actively used in the sensor concept of this thesis, e.g. for the detection of outliers. The answer is: The related geometrical conditions already contribute to the evaluation procedure. If a 3D point is evaluated from a camera image, the 'true' 3D point itself, plus all possibilities for outliers, are located on the same ray of sight, which is represented by the related epipolar line in the image of the second camera.

Commonly (e.g. in active or passive stereo), the epipolar geometry is exploited to *restrict the search range for corresponding points to one dimension*. Such an approach cannot be applied to the 3D movie camera. Point correspondences cannot be 'searched' by comparing the shapes of different signals as all signals in the camera images look the same.

10.3 Higher Pattern Frequencies with Rotated Cameras

This idea was contributed by *P. de Groot* from *Zygo Corporation*[1] (personal communication).

The discussions of Sect. 6.1 revealed that the line density of the 3D camera prototype is already close to its maximal possible value. For $N_L = 160$ projected (vertical) lines, the line distance d'_c in the picture of C_2 is about 7 pix for untilted surfaces, but may shrink to 3–4 pix at steep surface areas. Since a proper Gauß approximation requires a line width of at least 3 pix, the line density limit is reached.

The following consideration explains how this limit can be surpassed by a rotation of the lines in the camera image by 45°. The best way to achieve such a rotation in a concrete triangulation setup is to *rotate all cameras by* 45° without changing their position. This ensures that the sensor geometry and thus the triangulation angles between projection and cameras are not changed. The essential point of the idea is that the sub pixel precise maximum evaluation of the rotated lines is still performed row-wise (resp. column-wise for the other line direction). Then, two different types of line distances can be defined in the camera image: First, the distance d'_c between two lines, which is measured perpendicular to the line direction and defines the pattern frequency $\nu'_{lines} = 1/d'_c$. Second, the distance of two line signals in each row $d'_{eval} = \sqrt{2} \cdot d'_c$.

Figure 10.1 visualizes the situation. For the case of unrotated cameras in Fig. 10.1a it is obvious that $d'_c = d'_{eval}$. Both distances can not be smaller than about 7 pix to ensure correct maximum evaluation at steep surface parts. If the cameras are rotated (see Fig. 10.1b), a *decrease* of d'_c by a factor of $1/\sqrt{2}$ would *still ensure* correct maximum evaluation. Note that the horizontally sampled line profiles of the rotated lines are wider. Hence, the method only works for steep surface parts, if the width of the projected lines is also decreased—namely until the horizontally sampled line profiles of the rotated lines again have a width of about 3 pix.

Consequently, a rotation of the cameras allows for an *increase of the pattern frequency* by the factor $\sqrt{2}$. However, *the number of acquired 3D points remains unchanged!*

[1] www.zygo.com.

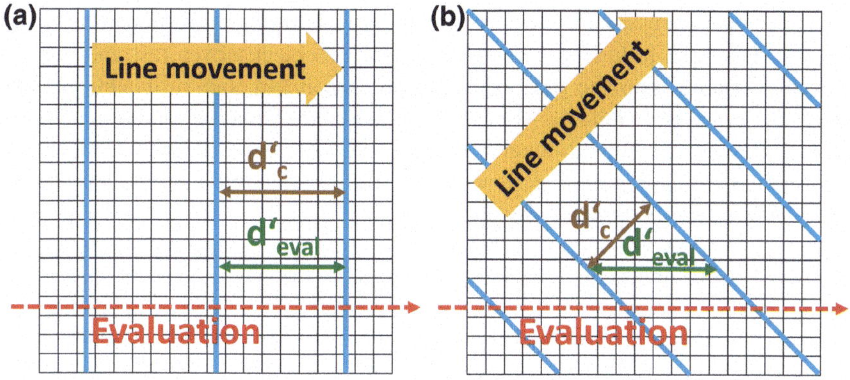

Fig. 10.1 Signal evaluation on the chip of an unrotated and a rotated camera. **a** Unrotated camera: Line distance and evaluation distance is equal. **b** Rotated camera: The line distance can be decreased in order to achieve the same evaluation distance as in (**a**)

According to the rotated sampling direction, the 3D data density *along each line* is decreased by the factor $1/\sqrt{2}$. This is not a significant drawback, since the 3D data density along each line is very high anyway.

Hence, the 3D points in the final dataset can be distributed more homogeneously if the cameras in the setup are rotated. This is a slight advantage compared to the unrotated case.

Important remark: A rotation of the cameras is not equivalent to a rotation of the projected pattern. If the sensor geometry is not changed, a rotation of the projected pattern would change the effective triangulation angles between pattern and cameras. This is understandable if the direction of the line movement during depth changes (yellow arrows in Fig. 10.1) is considered. If the *cameras* are rotated, the line movement is still perpendicular to the line direction. If the *projected pattern* is rotated, the line movement in Fig. 10.1b would also be in horizontal direction, which would lead to a different depth-sensitivities. This becomes especially important if a crossed line pattern is projected.

10.4 Movie Sequence of Ultra Fast Processes Within One Camera Frame

As discussed in several parts of this thesis, the time for the acquisition of a 3D model with the 3D movie camera is *independent* of the camera frame rate. If a proper flash illumination is applied, also 'slow' cameras are able to acquire 3D models without motion artifacts. It was also explained that, nevertheless, the acquisition of a *3D movie with high temporal resolution* relies on high camera frame rates.

This section introduces a consideration in order to surpass this problem. This is possible under the following conditions: First, the object moves fast and is considerably smaller than the field of view. Second, a fast and sufficiently bright strobe illumination is available. Moreover, a CCD-Chip displaying no motion related readout artifacts (e.g. rolling shutter artifact at CMOS-Chips) is assumed. Under these conditions, *several independent 3D models* of the moving object can be acquired at *different times within only one camera frame*.

The following example explains the procedure: Assuming, that the form-oszillation[2] of a flying tennis ball directly after the serve should be measured. The applied cameras assumed to have a moderate frame rate of 250 Hz. The 3D camera is placed so that the moving ball passes the field of view in horizontal direction. If the ball moves[3] at 75 $\frac{\mathrm{m}}{\mathrm{s}}$, it passes the horizontal field width of the 3D camera prototype ($\Delta X = 300$ mm) during exactly one camera frame of 4 ms. This means, that for an intermediate ball diameter of 68 mm, about 4 sequential, spatially separated 3D models of the deforming ball can be acquired within one camera frame, provided the illumination is triggered each millisecond. This would require exposure times far below 1 ms.

The basic idea for the measurement within one camera frame can be also applied on fast processes, which display no (or very low) spatial movement, e.g. explosions. Instead of the object, the 3D camera itself can be moved to 'sample' the measurement field at different illumination periods. The same effect can be achieved if a rotating mirror is applied.

Nevertheless, the considerations above tapped one of the basic problems of fast 3D metrology: The measurement of fast processes always requires *very bright light sources*. The sensor concept of the single-shot 3D movie camera may extenuate this problem, as a static illumination (which commonly reaches a higher luminance as switchable projectors) can be applied for the standard usage. However, the development of light sources with sufficient luminance is of great importance. Such light sources will enable many new applications for 3D sensors in the future.

References

1. C. Bräuer-Burchardt, P. Kühmstedt, G. Notni, Phase unwrapping using geometric constraints for high-speed fringe projection based 3D measurements. SPIE Opt. Metrol. **2013**, 878906–878906 (2013)
2. Z. Li, K. Zhong, Y. Li, X. Zhou, Y. Shi, Multiview phase shifting: a full-resolution and high-speed 3D measurement framework for arbitrary shape dynamic objects. Opt. Lett. **38**, 1389–1391 (2013)
3. K. Zhong, Z. Li, Y. Shi, C. Wang, Y. Lei, Fast phase measurement profilometry for arbitrary shape objects without phase unwrapping. Opt. Lasers Eng. **51**, 1213–1222 (2013)

[2]For a 2D slow motion see e.g. youtu.be/VHV1YbeznCo.

[3]The speed world record for a tennis serve is 263 $\frac{\mathrm{km}}{\mathrm{h}} = 73.1\ \frac{\mathrm{m}}{\mathrm{s}}$.

4. K. Zhong, Z. Li, X. Zhou, G. Zhan, X. Liu, Y. Shi, C. Wang, Real-time 3D shape measurement system with full temporal resolution and spatial resolution, in *Proceedings of SPIE 9013, Three-Dimensional Image Processing, Measurement (3DIPM), and Applications* (2014), p. 901309
5. R. Ishiyama, S. Sakamoto, J. Tajima, T. Okatani, K. Deguchi, Absolute phase measurements using geometric constraints between multiple cameras and projectors. Appl. Opt. **46**, 3528–3538 (2007)
6. K. Song, S. Hu, X. Wen, Y. Yan, Fast 3D shape measurement using Fourier transform profilometry without phase unwrapping. Opt. Lasers Eng. **84**, 74–81 (2016)
7. M. Young, E. Beeson, J. Davis, S. Rusinkiewicz, R. Ramamoorthi, Viewpoint-coded structured light, in *IEEE Computer Society Conference on Computer Vision and Pattern Recognition (CVPR)* (2007)
8. M. Kuch, Neue Indizierungsmethoden für die Multilinientriangulation. Master Thesis, University Erlangen-Nuremberg (2014)
9. R. Sagawa, Y. Ota, Y. Yagi, R. Furukawa, N. Asada, H. Kawasaki, Dense 3D reconstruction method using a single pattern for fast moving object, in *2009 IEEE 12th International Conference on Computer Vision*. (IEEE, 2009)

Chapter 11
Summary and Conclusion

In this thesis, a *novel optical sensor* for the *single-shot 3D metrology* of *moving objects* was *invented, investigated, developed* and *built* – the *'single-shot 3D movie camera'*. A major prerequisite was the understanding of the *physical limits* and *information limits* of *single-shot 3D sensing*.

The main problem to be solved was to widely eliminate the drawbacks of the single-shot principle 'line triangulation', while preserving all of its valuable features. The majority of single-shot principles compensate their lack of (temporal) information with an extensive spatial codification in order to solve the profound ambiguity problem. That was not an option, as the consumed space bandwidth is lost for a high feature resolution and data density of the acquired 3D model. The problem was solved by accessing a new source of information: *Perspective information* provided by *one or more additional synchronized cameras.*

Several variations of the basic idea have been discussed. Aside others, a statistic based method has been proposed to solve the ambiguity problem in 'Flying Triangulation'. For sensor setups that already include a second camera (e.g. for color texture acquisition) this method is purely software-based and can be applied without any hardware modifications. Thus, an implementation is possible for other line triangulation sensors as well.

To enable the desired single-shot 3D movie camera, the *density of acquired 3D data* per single-shot had to be *increased drastically*. This was achieved with another variation of the basic idea – the *'index-back-projection-approach'*. According to a deliberate geometrical arrangement of sensor components, this method *guarantees* unambiguous and dense 3D data within the whole effective measurement volume – a result achievable with only two cameras.

On the basis of the *'index-back-projection-approach'*, a prototype of the single-shot 3D movie camera was developed and built. The prototype acquires 3D data along 160 projected lines within a measurement volume of $\Delta X \times \Delta Y \times \Delta Z =$

F. Willomitzer, *Single-Shot 3D Sensing Close to Physical Limits and Information Limits*, Springer Theses,
https://doi.org/10.1007/978-3-030-10904-2_11

$300 \times 200 \times 100$ mm^3. According to the physical optimization of the sensor parameters, the acquired data display a precision better than $\delta z \leq 200$ µm inside the whole measurement volume. Along each (vertical) projected line a widely independent 3D point can be evaluated in each pixel-row on the chip. This guarantees - together with the small amount of bandwidth consumed for the sub pixel precise maximum evaluation in perpendicular line direction - for the resolution of high object frequencies in the final 3D model.

The acquired 3D data density was *again increased* (nearly by a factor two) by projecting a *crossed line pattern*. This was made possible *without the need of additional cameras* by applying a simple geometrical trick. Nearly 300,000 3D points can be acquired in each single-shot, if two 1-Megapixel cameras (1000×1000 pix) are applied. The prototype setup with effective detectors of 1024×682 pix respectively acquires nearly 190,000 3D points per single-shot. Nevertheless, the application of crossed lines requires a slight softening of the 'pure doctrine' of this thesis: The separation of line directions in the camera images introduces additional (not too severe) limitations regarding spatial bandwidth and precision – mainly at the crossing points.

In addition to the *optimization under physical aspects*, this thesis also attached importance to the *optimization under information theoretical aspects*. The so-called *'channel-efficiency'* was introduced as a measure of comparing different sensors. The single-shot 3D movie camera turned out to achieve a much higher channel efficiency than common single-shot sensors. The achieved values are *comparable* to the channel efficiency of very good PMT approaches.

In summary this finally leads to the conclusion that:

> The *single-shot 3D movie camera* acquires 3D data ***close to the limits of physics and information theory***, achievable with the applied hardware.

Only in case this requirement is met, it is worthwhile investing in better hardware in order to further reduce technical restrictions. The prototype setup, only using standard low-cost components and tailored to the measurement of human faces, is considered as a *proof of the principle*. A single-shot 3D movie camera containing high speed cameras with high resolution and precise optics as well as a bright projector, will display *unprecedented features*. It will enable the precise 3D measurement of *very fast processes* such as deforming car bodies or inflating airbags during crash tests.

Compared to existing multi-shot approaches, which are also able to measure such processes by applying immensely fast cameras and projectors, the single-shot 3D movie camera will naturally have two advantages: First, the time required for the acquisition of a single 3D model is *solely dependent on the exposure time* and thus independent of the camera frame rate, meaning the system is *much more robust against motion artifacts*. Second, assuming equal hardware components, the *temporal resolution* of '3D movies' will *always be better* by a factor of F (subsequently acquired frames in the multi-shot principle) according to the single-shot ability.

Consequently, the vast potential of the *single-shot 3D movie camera* is *far from being fully exploited* yet.

About the Author

Florian Willomitzer is currently working as a Postdoctoral Fellow at Northwestern University, US. He graduated from the University of Erlangen-Nuremberg, Germany, where he received his Ph.D. degree with honors ('summa cum laude') in 2017. At Erlangen, Dr. Willomitzer worked in the group 'Optical Sensing, Metrology and Inspection', led by Prof. Gerd Häusler. During his doctoral studies he investigated physical and information theoretical limits of optical 3D-sensing and implemented sensors that operate close to these limits. This work is distilled in the present Dissertation. Concurrent to his activity at the Erlangen University, Dr. Willomitzer was a freelancer in the research group's spin-off company '3D-Shape GmbH' and worked as a high school part time teacher for physics.

At Northwestern University, Dr. Willomitzer's research is focused on emerging optical measurement principles to obtain non-line-of-sight information, novel techniques to overcome traditional resolution limitations and dynamic range restrictions in 3D and 2D imaging, and the implementation of high-precision metrology methods using low-cost mobile handheld devices.

F. Willomitzer, *Single-Shot 3D Sensing Close to Physical Limits and Information Limits*, Springer Theses,
https://doi.org/10.1007/978-3-030-10904-2

Appendix A
Full List of Specifications for the Prototype Setup

Table A.1 Full list of specifications for the prototype setup

Measurement field[a,b]	$\Delta X \times \Delta Y$	300×200 mm^2
Effective measurement range	$\Delta Z_e \leq \Delta Z_1$	100 mm
Measurement uncertainty	δz	≤ 200 μm
Stand-off distance[a,c]	w	~ 500 mm
Number of active pixels[b]	$N_x \times N_y$	1024×682
Pixel size	$(d'_{pix})^2$	$(4.65$ μm$)^2$
Frame rate	r_f	30 Hz
Numerical observation aperture[a]	$\sin u_{obs}$	0.00225
Numerical illumination aperture[a]	$\sin u_{ill}$	0.0075
Mean illumination wavelength	λ	550 nm
Cameras	*AVT Guppy F-080 B/C* [1]	
Objectives	*Fujinon HF9HA-1B* [2]	
Projector	*Acer K132* [3]	
Setup with unidirectional lines		
Triangulation angle[a] sensor 1	θ_1	$\sim 1°$
Triangulation angle[a] sensor 2	θ_2	$\sim 9°$
Number of projected lines	N_L	≤ 160
3D Points per single-shot	N_{3D}	$\lesssim 100{,}000$
Channel efficiency	η	$\leq 11\%$
Setup with crossed lines		
Triangulation angle[a] sensor 1v and 2h	$\theta_{1,v}, \theta_{2,h}$	$\sim 1°$
Triangulation angle[a] sensor 2v and 1h	$\theta_{2,v}, \theta_{1,h}$	$\sim 9°$
Number of projected vertical lines	$N_{L,v}$	≤ 160
Number of projected horizontal lines	$N_{L,h}$	≤ 100
3D points per single-shot	N_{3D}	$\lesssim 190{,}000$
Channel efficiency	η	$\leq 20.3\%$

[a] in the middle of the measurement range
[b] total overlap of projection and observation
[c] measured from the first objective lens

F. Willomitzer, *Single-Shot 3D Sensing Close to Physical Limits and Information Limits*, Springer Theses,
https://doi.org/10.1007/978-3-030-10904-2

References

1. AlliedVision, "Guppy F-080B/C Datasheet" (2017). www.alliedvision.com/en/products/cameras/detail/Guppy/F-080/action/pdf.html. Accessed 15 May 2017
2. Fujinon, "HF9HA-1B Datasheet" (2017). www.fujifilmusa.com/shared/bin/1.5Mega-Pixel2.31.2.pdf. Accessed 15 May 2017
3. Acer, "Projector K132 Datasheet" (2017). www.tinyurl.com/AcerK132-Details-pdf. Accessed 15 May 2017

The manufacturer's authorised representative in the EU is Springer Nature Customer Service Centre GmbH, Europaplatz 3, 69115 Heidelberg, Germany. If you have any concerns regarding our products, please contact ProductSafety@springernature.com

Printed and bound by CPI Group (UK) Ltd, Croydon, CR0 4YY

12/08/2026

02194517-0001